T0137349

Studies in Big Data

Volume 35

Series editor

Janusz Kacprzyk, Polish Academy of Sciences, Warsaw, Poland
e-mail: kacprzyk@ibspan.waw.pl

The series "Studies in Big Data" (SBD) publishes new developments and advances in the various areas of Big Data- quickly and with a high quality. The intent is to cover the theory, research, development, and applications of Big Data, as embedded in the fields of engineering, computer science, physics, economics and life sciences. The books of the series refer to the analysis and understanding of large, complex, and/or distributed data sets generated from recent digital sources coming from sensors or other physical instruments as well as simulations, crowd sourcing, social networks or other internet transactions, such as emails or video click streams and other. The series contains monographs, lecture notes and edited volumes in Big Data spanning the areas of computational intelligence incl. neural networks, evolutionary computation, soft computing, fuzzy systems, as well as artificial intelligence, data mining, modern statistics and Operations research, as well as self-organizing systems. Of particular value to both the contributors and the readership are the short publication timeframe and the world-wide distribution, which enable both wide and rapid dissemination of research output.

More information about this series at http://www.springer.com/series/11970

Han Liu · Mihaela Cocea

Granular Computing Based Machine Learning

A Big Data Processing Approach

 Springer

Han Liu
School of Computer Science
  and Informatics
Cardiff University
Cardiff
UK

Mihaela Cocea
School of Computing
University of Portsmouth
Portsmouth
UK

ISSN 2197-6503   ISSN 2197-6511   (electronic)
Studies in Big Data
ISBN 978-3-319-88884-2   ISBN 978-3-319-70058-8   (eBook)
https://doi.org/10.1007/978-3-319-70058-8

Printed on acid-free paper

This Springer imprint is published by Springer Nature
The registered company is Springer International Publishing AG
The registered company address is: Gewerbestrasse 11, 6330 Cham, Switzerland

It's not who has the best algorithm that wins. It's who has the most data.

—Andrew Ng

Preface

The ideas introduced in this book explore the relationships among big data, machine learning and granular computing. In many studies, machine learning has been considered as a powerful tool of big data processing. The relationship between big data and machine learning is very similar to the relationship between resources and human learning. In this context, people can learn from resources to deal with new matters. Similarly, machines can learn from big data to resolve new problems. However, due to the vast and rapid increase in the size of data, learning tasks have become increasingly more complex. In this context, traditional machine learning has been too shallow to deal with big data sufficiently, so granular computing concepts are used in this book to advance machine learning towards the shift from shallow learning to deep learning (in its broader sense).

The focus of this book is on the development and evaluation of granular computing based machine learning approaches in terms of classification accuracy. In this context, the authors consider traditional machine learning to be of single-granularity and the proposal of granular computing based machine learning is aimed at turning single-granularity learning into multi-granularity learning. In particular, the authors proposed the following transformations: (a) supervised learning to semi-supervised learning, (b) heuristic learning to semi-heuristic learning, (c) single-task learning to multi-task learning, (d) discriminative learning to generative learning and (e) random data partitioning to semi-random data partitioning. In addition, the authors also explore how to achieve in-depth evaluation of attribute-value pairs towards induction of high-quality rules, in the setting of multi-granularity learning.

Multi-granularity learning is not only a scientific proposal to address issues of traditional machine learning, but also an indication of philosophical inspiration from real-life problems. For example, it is assumed in traditional machine learning that different classes are mutually exclusive and each instance is clear-cut. However, this assumption does not always hold in reality, e.g. the same movie may belong to more than one category or the same book may belong to more than one subject. This indicates that there could be specific relationships between different classes, such as mutual exclusion, correlation and mutual independence, which has inspired the authors to adopt the concept of relationships between granules. In this context, each class is viewed as a granule, and generative learning needs to be adopted instead of discriminative learning, if the relationship between classes is not mutual exclusion.

On the other hand, in the context of computer science, correlation is also referred to as association, which is considered as a horizontal relationship between classes. Also, different classes may involve hierarchical relationships such as inheritance and aggregation. In other words, one class can be specialized or decomposed into several sub-classes and several classes may be generalized or aggregated into a super-class. In this book, the authors define that classes that involve a horizontal relationship are viewed as granules located in the same level of granularity and that classes that involve hierarchical relationships are viewed as granules located in different levels of granularity.

Based on the above definition, if all the classes are in the same level of granularity, the learning task could simply be undertaken in the setting of single-granularity learning, unless the task is complex. In particular, it would be considered the case of single-task learning, if the learning task is just aimed to discriminate one class from the other classes towards classifying a clear-cut instance. Otherwise, multi-task learning needs to be undertaken, if the learning task is aimed at judging independently on each class in terms of the membership or non-membership of an instance to the class. In contrast, if these classes are in different levels of granularity, it would usually be necessary to undertake multi-granularity learning.

Multi-granularity learning is also inspired philosophically from the examples of human learning and military training. In the context of human learning, a student typically needs to take a course that involves modules in different levels of a degree. In other words, a student needs to learn and pass all the basic modules in one level, such that the student can progress to the next level towards learning and passing more advanced modules. In the setting of granular computing, each module of a course is viewed as a granule and each level of a degree is viewed as a level of granularity.

In the context of military training, a solider normally needs to be with a particular unit at one of the military levels, such as squad, platoon, company, battalion, regiment, division and corps. In other words, military training could be organized at different scales, which needs to consider units of different levels as basic units, e.g. a company can be considered as a basic unit in a small scale of training, whereas a regiment needs to be considered as a basic unit in a large scale of training. In the setting of granular computing, each unit is viewed as a granule and each military level is viewed as a level of granularity.

For each of the transformations discussed in this book (e.g. supervised to semi-supervised learning, heuristic to semi-heuristic learning), the granular framework is clearly outlined. In addition, two case studies are presented, one on biomedical data and one on sentiment analysis.

Cardiff, UK   Han Liu
Portsmouth, UK   Mihaela Cocea

Acknowledgements

The first author would like to thank the University of Portsmouth for appointing him to a research associate position to conduct the research activities leading to the results disseminated in this book. Special thanks must go to his parents Wenle Liu and Chunlan Xie as well as his brother Zhu Liu for the financial support during his academic studies in the past as well as the spiritual support and encouragement for his embarking on a research career in recent years. In addition, the first author would also like to thank his best friend Yuqian Zou for the continuous support and encouragement during his recent research career that have facilitated significantly his involvement in the writing process for this book.

The authors would like to acknowledge support for the research presented in this book through the Research Development Fund at the University of Portsmouth. The authors would like to thank the editor of the series *Studies in Big Data*, Prof. Janusz Kacprzyk, and the executive editor for this series, Dr. Thomas Ditzinger, for the support and encouragement for disseminating our research work through this book. The authors also give thanks to Prof. Witold Pedrycz from University of Alberta and Prof. Shyi-Ming Chen from National Taiwan University of Science and Technology for the feedback received on the research presented in this book, as well as the support and advice on career development.

Contents

Acronyms

Bagging	Bootstrap Aggregating
BOS	Bag-of-Sentences
BOW	Bag-of-Words
BP	Back-Propagation
CART	Classification and Regression Trees
CI	Computational Intelligence
GA	Genetic Algorithm
IEBRG	Information Entropy Based Rule Generation
KNN	K Nearest Neighbours
NB	Naive Bayes
NLP	Natural Language Processing
PNN	Probabilistic Neural Networks
POS	Part-of-Speech
RF	Random Forests
SVM	Support Vector Machine
TDIDT	Top-Down Induction of Decision Trees
UCI	University of California, Irvine

Chapter 1
Introduction

Abstract In this chapter, we will introduce the background of big data, in terms of five Vs, namely volume, velocity, variety, veracity and variability. The concepts of traditional data science are then explored to show the value of data. Furthermore, the concepts of machine learning and granular computing are provided in the context of intelligent data processing. Finally, the main contents of each of the following chapters are outlined.

1.1 Background of Big Data

Big data can generally be characterized by 5Vs Volume, Velocity, Variety, Veracity and Variability. In particular, volume generally reflects the space required to store data. Velocity reflects the speed of data transmission and processing, i.e. how effectively and efficiently real-time data is collected and processed on the platform of cloud computing. Variety reflects the type of data, i.e. data can be structured or unstructured and can also be in different forms such as text, image, audio and video. Veracity reflects the degree to which data can be trusted. Variability reflects the dissimilarity between different instances in a data set. More details on big data can be found in [1–3].

In many studies, machine learning has been considered as a powerful tool of big data processing. As introduced in [4], the relationship between big data and machine learning is very similar to the relationship between resources and human learning. In this context, people can learn from resources to deal with new matters. Similarly, machines can learn from big data to resolve new problems. More details on big data processing by machine learning can be found in [5, 6].

Machine learning is also regarded as one of the main approaches of computational intelligence (CI) [7] for which big data is considered as an environment [3]. In general, CI encompasses a set of nature or biology inspired computational approaches such as artificial neural networks, fuzzy systems and evolutionary computation. In particular, artificial neural networks are biologically inspired to simulate the human brain in terms of learning through experience. Also, fuzzy systems involve using fuzzy logic, which enables computers to understand natural languages [8]. Moreover,

© Springer International Publishing AG 2018
H. Liu and M. Cocea, *Granular Computing Based Machine Learning*,
Studies in Big Data 35, https://doi.org/10.1007/978-3-319-70058-8_1

evolutionary computation works based on the process of nature selection, learning theory and probabilistic methods, which helps with uncertainty handling [7]. As stated in [9], learning theories, which help understand how cognitive, emotional and environmental effects and experiences are processed in the context of psychology, can help make predictions on the basis of previous experience in the context of machine learning. From this point of view, machine learning is naturally inspired by human learning and would thus be considered as a nature-inspired approach. In addition, most machine learning methods involve employing heuristics of computational intelligence, such as probabilistic measures, fuzziness and fitness, towards optimal learning.

In machine learning context, learning algorithms are typically evaluated in four dimensions, namely accuracy, efficiency, interpretability and stability, following the concepts of computational intelligence. These four dimensions can be strongly related to veracity, volume, variety and variability, respectively.

Veracity reflects the degree to which data can be trusted as mentioned above. In practice, data needs to be transformed into information or knowledge for people to use. From this point of view, the accuracy of information or knowledge discovered from data can be highly impacted by the quality of the data and thus is an effective way of evaluation against the degree of trust.

Volume reflects the size of data. In the areas of machine learning and statistics, the data size can be estimated through the product of data dimensionality and sample size [10]. Increase of data dimensionality or sample size can usually increase the computational costs of machine learning tasks. Therefore, evaluation of the volume for particular data is highly related to estimation of memory usage for data processing by machine learning methods.

Variety reflects the format of data, i.e. data types and representation. Typical data types include integer, real, Boolean, string, nominal and ordinal [11]. In machine learning and statistics, data types can be simply divided into two categories: discrete and continuous. On the other hand, data can be represented in different forms, e.g. text, graph and tables. All the differences mentioned above in terms of data format can impact on the interpretability of models learned from data.

Variability reflects the dissimilarity between different instances in a data set. In machine learning, the performance of learning algorithms can appear to be highly unstable due to change of data samples, especially when the data instances are highly dissimilar to each other. Therefore, the stability of a learning algorithm can be highly impacted by data variability.

The above four aspects (accuracy, efficiency, interpretability and stability) are also impacted greatly by the selection of different machine learning algorithms. For example, data usually needs to be pre-processed by particular algorithms prior to the training stage, which leads to a particular level of impact on data modelling. Also, inappropriate sampling of training and test data can also lead to building a poor model and biased estimation of accuracy, respectively. Further, different learning algorithms can usually lead to different quality of models learned from the same training data. In addition, in the context of online learning, velocity, which is related to the learning speed of an algorithm, is an important impact factor for data streams to be processed effectively and efficiently. However, this chapter focuses on offline learning, which analyses in depth how the nature of learning algorithms is related to the nature of static data.

1.2 Concepts of Data Science

In the context of data science, data, information and knowledge are considered as three main concepts. In particular, it can be viewed that data can be transformed into information and then be transformed into knowledge, as illustrated in Fig. 1.1.

From computer science perspective, data is viewed as figures and symbols, which are aimed at passing messages between senders and receivers. In contrast, information is viewed as contextualized, organized and interpretable pictures resulting from data. From this point of view, information is generally more interpretable than data. For example, Alice sends an encrypted message to Bob. In this context, as the message is encrypted and thus is not understandable, the message can only be considered as data rather than information. In order to understand the message, Bob needs to decrypt it. Following the decryption, the message can then be considered as information.

In comparison with knowledge, information typically retains high uncertainty in terms of the degree of truth. In other words, knowledge is considered as a highly trusted source of information. For example, in military, when an intelligence is received, it is necessary to verify through different sources towards confirming the truth of the intelligence. In this context, an intelligence can only be considered as information until the intelligence has been confirmed in terms of its truth. In other words, an intelligence can be considered as knowledge following the confirmation of the truth of this intelligence.

From machine learning perspective, the training stage is aimed at transforming data into information. In other words, training data is used for algorithms to learn models. In this context, the models are considered as information transformed from the training data. Moreover, the testing stage is aimed at transforming information

Fig. 1.1 Transformation from data through information to knowledge

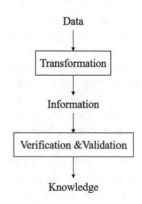

into knowledge. In other words, test data is used to evaluate the degree to which the models are trusted. In this context, the models can only be considered as information until it has been confirmed that the models are highly trusted. In other words, the models can be used as knowledge, following the evaluation that confirms that the models are trustworthy. More details on machine learning are provided in Sect. 1.3.

1.3   Machine Learning

Machine learning is a branch of artificial intelligence and involves two stages: training and testing [12]. The first stage aims to learn something from known properties by using learning algorithms and the second stage aims to make predictions on unknown properties by using the knowledge learned in the first stage. From this point of view, training and testing are also referred to as learning and prediction, respectively. In practice, a machine learning task is aimed at building a model, which is further used to make predictions, through the use of learning algorithms. Therefore, this task is usually referred to as predictive modelling.

Machine learning could be divided into two special types: supervised learning and unsupervised learning [12], in terms of the form of learning. Supervised learning means learning with a teacher, because all instances from a training set are labelled, which makes the learning outcomes very explicit. In other words, supervised learning is naturally inspired by student learning with the supervision of teachers. In practice, the aim of this type of learning is to build a model by learning from labelled data and then to make predictions on other unlabeled instances with regard to the value of a predicted attribute. The predicted value of an attribute could be either discrete or continuous. Therefore, supervised learning could be involved in both classification and regression tasks for categorical prediction and numerical prediction, respectively. In contrast, unsupervised learning means learning without a teacher. This is because all instances from a training set are unlabeled and thus the learning outcomes are not explicit. In other words, unsupervised learning is naturally inspired by student learning without being supervised. In practice, the aim of this type of learning is to discover previously unknown patterns from data sets. It includes association and clustering. The former aims to identify correlations between attributes, whereas the latter aims to group objects on the basis of their similarity to each other.

According to [12], machine learning algorithms can be put into several categories: decision tree learning, rule learning, instance based learning, Bayesian learning, perceptron learning and ensemble learning. All of these learning algorithms show the characteristic of nature inspiration.

Both decision tree learning and rule learning aim to learn a set of rules on an inductive basis. However, the difference between the two types of learning is that the former generates rules in the form of a decision tree and the latter generates if-then rules directly from training instances [1, 13, 14]. The above difference is mainly due to the fact that the former follows the divide and conquer approach [15] and the latter follows the separate and conquer approach [16]. In particular, the divide and conquer approach is naturally similar to the top-down approach of student learning, such as dividing a textbook into several levels: parts, chapters, sections and subsections. The separate and conquer approach is naturally similar to the iterative approach of student learning, which means by reading through an entire material in the first iteration and then focusing on more important parts of the material for deeper understanding in the subsequent iterations.

Instance based learning generally involves predicting test instances on the basis of their similarity to the training instances, such as K nearest neighbors (KNN) [17]. This type of learning is also referred to as lazy learning, due to the fact that it does not aim to learn in depth to gain some pattern from data but just to make as many correct predictions as possible [10]. In other words, this type of learning is naturally similar to the exam centered approach of student learning, which means that students mainly aim to answer correctly the exam questions without deep understanding of knowledge.

Bayesian learning essentially employs the Bayes theorem [18]. In particular, this type of learning is based on the assumption that all the input attributes are totally independent of each other. In this context, each attribute-value pair would be independently correlated to each of the possible classes, which means that a posterior probability is provided between the attribute-value pair and the class. A popular method of Bayesian learning is Naive Bayes (NB) [19]. This type of learning is naturally similar to the prior-knowledge based approach of human reasoning, which means that people make decisions, and apply reasoning and judgments based on the knowledge they obtained before, towards having the most confident choice.

Perceptron learning aims to build a neural network topology that consists of a number of layers and that has a number of nodes, each of which represents a perceptron. Some popular algorithms include back-propagation (BP) [11] and probabilistic neural networks (PNN) [12]. This type of learning is biology inspired as stated in Sect. 1.1. Ensemble learning generally aims to combine different learning algorithms in the training stage or computational models in the testing stage towards improvement of overall accuracy of predictions. Some popular approaches of ensemble learning include bagging [20] and boosting [21]. This type of learning is naturally similar to the approach of group learning for students to collaborate on a group assignment.

In terms of evaluating a machine learning task, there are generally two main approaches: cross-validation and split of data into a training set and a test set. Cross-validation generally means to split a data set into n disjoint subsets. In this context, there would be n iterations in total for the evaluation, while at each iteration a subset is used for testing and the other $n - 1$ subsets are used for training. In other words, each of the n subsets is in turn used as the test set at one of the n iterations, while the rest of the subsets are used together as the training set. In laboratory research, 10-fold cross-validation is used more popularly, i.e. the original data set is split into 10 subsets. Cross-validation is generally more expensive in terms of computational cost. Therefore, researchers sometimes instead choose to take the approach of splitting a data set into a training set and a test set in a specific ratio, e.g. 70% of the data is used as the training set and the rest of the data is used as the test set. This data split can be done randomly or in a fixed way. However, due to the presence of uncertainty in data, the random split of data is more popular for researchers in machine learning or similar areas.

In this book, new perspectives of the two approaches of evaluating machine learning tasks are used in Chap. 8. In particular, cross-validation is used towards measuring effectively the learnability of an algorithm, i.e. the extent to which the algorithm is suitable to build a confident model on the provided training data. This is in order to help employ appropriately the suitable learning algorithms for building predictive models on the basis of existing data. The other approach for splitting a data set into a training set and a test set is adopted towards learning a model that covers highly complete patterns from the training data and evaluating the model accuracy using highly similar but different instances from the test data. This is in order to ensure the model accuracy evaluated by using the test data is trustworthy. Details on the use of the new perspectives are presented in Chap. 8.

1.4 Granular Computing

Granular computing has been an increasingly popular approach for in-depth processing of information. It is aimed at structural thinking at the philosophical level, as well as at structural problem solving at the practical level [22]. In general, granular computing involves two operations, namely, granulation and organization. The former operation means to decompose a whole into several parts, whereas the latter operation means to integrate several parts into a whole. From computer science perspective, granulation corresponds to the top-down approach and organization corresponds to the bottom-up approach. The nature of granular computing involves two commonly used concepts, namely, granule and granularity.

In the context of information granule, a granule is defined as "a small particle; especially, one of numerous particles forming a larger unit", according to the Merriam-Websters Dictionary [23]. In practice, there have been various examples of granules in broad application areas.

In the setting of set theory, a set of any formalism can be viewed as a granule, since a set is a collection of elements. In this context, each element is viewed as a particle. Different formalisms of sets include deterministic sets [1], probabilistic sets [24], fuzzy sets [25] and rough sets [26].

In the area of computer science, a granule can act as a class due to the fact that a class is a group of objects which are highly similar to each other. An object can also be viewed as a granule, since each object involves a number of attributes, each of which is considered as a particle. Moreover, a granule can also act as a cluster due to the fact that clustering is another way of grouping objects.

In the area of natural languages, a document could be organized in different forms of text units, such as chapters, sections, paragraphs, sentences and words. In this context, each form of text unit can be viewed as a special type of granule. Moreover, each word is viewed as the finest granule due to the fact that a word consists of letters, each of which is viewed as a particle [27].

The concept of information granules is also popularly involved in other application areas, such as image processing, machine learning and rule based systems. More details on information granules can be found in [2, 26, 28, 29].

In the context of information granularity, information granules can be located in different levels of granularity. In set theory, a set S may have several subsets $(S_1, S_2, ...S_n)$ and each subset may also have several sub-subsets $(S_{1.1}, S_{1.2}, ...S_{1.m}, ...$ $S_{n.1}, S_{n.2}, ...S_{n.m})$. In this context, the set S is a granule in the top level of granularity, the subsets $(S_1, S_2, ...S_n)$ are in the middle level of granularity, and the sub-subsets $(S_{1.1}, S_{1.1}, ...S_{1.m}, ...S_{n.1}, S_{n.2}, ...S_{n.m})$ are in the bottom level of granularity. In computer science, a class can be specialized into several sub-classes through information granulation. Also, sub-classes can be generalized into a super class through information organization.

In natural language processing (NLP), a document can be managed in a granular structure. In particular, the complexity of a text instance (granule) can be reduced through top-down decomposition (granulation) in order to enable text units (granules) in different levels of granularity (such as paragraphs, sentences, and words) to be processed separately. Also, the outcomes for processing text units in the same level of granularity can be combined through bottom-up aggregation (organization) towards deriving the outcome for processing larger text units in a higher level of granularity.

In real applications, techniques of granular computing have been involved very often in other popular areas, such as artificial intelligence [22, 30, 31], computational intelligence [22, 32–34], and machine learning [35–38].

Furthermore, ensemble learning is also a subject that involves applications of granular computing concepts [37]. In particular, ensemble learning approaches, such as Bagging, involve information granulation through decomposing a training set into a number of overlapping samples and combining the predictions made from different classifiers towards classifying a test instance; a similar perspective has also been stressed and discussed in [39]. Chapter 6 will show how granular computing concepts can be used towards more effective partitioning of data for machine learning experimentation.

1.5 Chapters Overview

This book consists of nine main chapters, namely, traditional machine learning, semi-supervised learning through machine based labelling, nature inspired semi-heuristic learning, fuzzy classification through generative multi-task learning multi-granularity semi-random data partitioning, multi-granularity rule learning, case studies and conclusion.

In Chap. 2, we describe the characteristics of traditional machine learning. In particular, we describe traditional machine learning from several perspectives: supervised learning, heuristic learning, single task learning, discriminative learning, and random data partitioning. Also, we identify general issues of traditional machine learning and analyze their impact in the context of big data.

In Chap. 3, based on the issues of supervised learning identified in Chap. 2, we present the proposal of semi-supervised learning towards addressing the issues especially in the context of big data processing. In particular, we describe the concepts of semi-supervised learning and illustrate the proposed framework of learning in the context of granular computing.

In Chap. 4, based on the issues of heuristic learning identified in Chap. 2, we present the proposal of semi-heuristic learning, which is naturally and biologically inspired, towards addressing the issues especially in the context of big data processing. In particular, we describe the concepts of semi-heuristic learning and illustrate the nature and biology inspired framework of learning in the context of granular computing.

In Chap. 5, based on the issues of single-task learning and discriminative learning, we present the proposal of fuzzy classification through generative multi-task learning, towards addressing the issues especially in the context of big data processing. In particular, we describe the concepts of both multi-task learning and generative learning, and illustrate the framework of fuzzy generative multi-task learning.

In Chap. 6, based on the issues of random data partitioning identified in Chap. 2, we present the proposal of semi-random data partitioning. In particular, we describe the concepts of data partitioning and illustrate the multi-granularity framework of semi-random data partitioning.

In Chap. 7, based on the issues of rule learning identified in machine learning literature, we presents the proposal of multi-granularity rule learning. In particular, we describe the concepts of rule learning and illustrate the multi-granularity framework towards in-depth learning of rules.

In Chap. 8, we present several case studies of big data by using biomedical data and sentiment data, respectively, in order to show how big data processing can be advanced through the shift from traditional machine learning to granular computing based machine learning.

In Chap. 9, we will stress the theoretical significance, practical importance, methodological impact and philosophical aspects of granular computing based machine learning, and several further directions are suggested towards machine learning advances to fit the needs of modern industries.

References

1. Liu, H., A. Gegov, and M. Cocea. 2016. *Rule based systems for big data: a machine learning approach*. Switzerland: Springer.
2. Pedrycz, W., and S.-M. Chen. 2015. *Information granularity, big data, and computational intelligence*. Heidelberg: Springer.
3. Pedrycz, W., and S. M. Chen. 2017. *Data science and big data: an environment of computational intelligence*. Heidelberg: Springer.
4. P. Levine. Machine learning + big data. http://a16z.com/2015/01/22/machine-learning-big-data/

5. Liu, H., A. Gegov, and M. Cocea. 2017. Unified framework for control of machine learning tasks towards effective and efficient processing of big data. *In Data Science and Big Data: An Environment of Computational Intelligence*, 123–140. Switzerland: Springer.

6. Wu, X., X. Zhu, G.-Q. Wu, and W. Ding. 2014. Data mining with big data. *IEEE Transactions on Knowledge and Data Engineering* 26 (1): 97–107.

7. Siddique, N., and H. Adeli. 2013. *Computational intelligence: synergies of fuzzy logic, neural networks and evolutionary computing*. New Jersey: Wiley.

8. Rutkowski, L. 2008. *Computational intelligence: methods and techniques*. Heidelberg: Springer.

9. J. Worrell. 2014. Computational learning theory: 2014–2015. https://www.cs.ox.ac.uk/teaching/courses/2014-2015/clt/

10. H. Liu, M. Cocea, and A. Gegov. 2016. Interpretability of computational models for sentiment analysis. In *Sentiment Analysis and Ontology Engineering: An Environment of Computational Intelligence*, W. Pedrycz and S.-M. Chen, eds., vol. 639, 199–220.

11. Tan, P.-N., M. Steinbach, and V. Kumar. 2006. *Introduction to Data Mining*. New Jersey: Pearson Education.

12. Mitchell, T. 1997. *Machine Learning*. New York: McGraw Hill.

13. H. Liu, A. Gegov, and M. Cocea. 2015. Network based rule representation for knowledge discovery and predictive modelling. In *IEEE International Conference on Fuzzy Systems*, Istanbul, Turkey, 2–5 August 2015, 1–8.

14. H. Liu, A. Gegov, and F. Stahl. 2014. Categorization and construction of rule based systems. In *15th International Conference on Engineering Applications of Neural Networks*, Sofia, Bulgaria, 5–7 September 2014, 183–194.

15. Quinlan, R.J. 1986. Induction of decision trees. *Machine Learning* 1 (1): 81–106.

16. Furnkranz, J. 1999. Separate-and-conquer rule learning. *Artificial Intelligence Review* 13: 3–54.

17. J. Zhang. 1992. Selecting typical instances in instance-based learning. In *Proceedings of the Ninth International Workshop on Machine Learning*, Aberdeen, United Kingdom, 1–3 July 1992, 470–479.

18. Hazewinkel, M. 2001. *Encyclopedia of Mathematics*. London: Springer.

19. I. Rish. 2001. An empirical study of the naive bayes classifier. *IJCAI 2001 Workshop On Empirical Methods In Artificial Intelligence*, 3 (22), 41–46.

20. Breiman, L. 1996. Bagging predictors. *Machine Learning* 24 (2): 123–140.

21. Y. Freund and R. E. Schapire. 1996. Experiments with a new boosting algorithm. In *Machine Learning: Proceedings of the Thirteenth International Conference*, Bari, Italy, 3–6 July 1996, 148–156.

22. Y. Yao. 2005. Perspectives of granular computing. In *Proceedings of 2005 IEEE International Conference on Granular Computing*, Beijing, China, 25–27 July 2005, 85–90.

23. Merriam-Webster. 2016. Merriam-websters dictionary. http://www.merriam-webster.com/

24. Liu, H., A. Gegov, and M. Cocea. 2016. Rule based systems: A granular computing perspective. *Granular Computing* 1 (4): 259–274.

25. Zadeh, L. 2015. Fuzzy logic: A personal perspective. *Fuzzy Sets and Systems* 281: 4–20.

26. Pedrycz, W. 2011. Information granules and their use in schemes of knowledge management. *Scientia Iranica* 18 (3): 602–610.

27. H. Liu and M. Cocea. Fuzzy information granulation towards interpretable sentiment analysis. *Granular Computing*, 3 (1), In press.

28. Pedrycz, W., and S.-M. Chen. 2011. *Granular computing and intelligent systems: design with information granules of higher order and higher type*. Heidelberg: Springer.

29. Pedrycz, W., and S.-M. Chen. 2015. *Granular computing and decision-making: interactive and iterative approaches*. Heidelberg: Springer.

30. Wilke, G., and E. Portmann. 2016. Granular computing as a basis of humandata interaction: a cognitive cities use case. *Granular Computing* 1 (3): 181–197.

31. Skowron, A., A. Jankowski, and S. Dutta. 2016. Interactive granular computing. *Granular Computing* 1 (2): 95–113.

32. Dubois, D., and H. Prade. 2016. Bridging gaps between several forms of granular computing. *Granular Computing* 1 (2): 115–126.
33. Kreinovich, V. 2016. Solving equations (and systems of equations) under uncertainty: how different practical problems lead to different mathematical and computational formulations. *Granular Computing* 1 (3): 171–179.
34. Livi, L., and A. Sadeghian. 2016. Granular computing, computational intelligence, and the analysis of non-geometric input spaces. *Granular Computing* 1 (1): 13–20.
35. Min, F., and J. Xu. 2016. Semi-greedy heuristics for feature selection with test cost constraints. *Granular Computing* 1 (3): 199–211.
36. Peters, G., and R. Weber. 2016. DCC: A framework for dynamic granular clustering. *Granular Computing* 1 (1): 1–11.
37. Liu, H., and M. Cocea. 2017. Granular computing based approach for classification towards reduction of bias in ensemble learning. *Granular Computing* 2 (3): 131–139.
38. Antonelli, M., P. Ducange, B. Lazzerini, and F. Marcelloni. 2016. Multi-objective evolutionary design of granular rule-based classifiers. *Granular Computing* 1 (1): 37–58.
39. H. Hu and Z. Shi. 2009. Machine learning as granular computing. In *IEEE International Conference on Granular Computing*, Nanchang, Beijing, 17–19 August 2009, 229–234.

Chapter 2
Traditional Machine Learning

Abstract In this chapter, we describe the concepts of traditional machine learning. In particular, we introduce the key features of supervised learning, heuristic learning, discriminative learning, single-task learning and random data partitioning. We also identify general issues of traditional machine learning, and discuss how traditional learning approaches can be impacted due to the presence of big data.

2.1 Supervised Learning

Supervised learning means learning with a teacher, i.e. data used in the training stage must be fully labeled by experts. In practice, supervised learning can be involved in classification and regression tasks. The main difference between classification and regression is that the output attribute must be discrete for the former type of tasks, whereas the output attribute must be continuous for the latter type of tasks. From this point of view, classification is also known as categorical prediction. Similarly, regression is also known as numerical prediction.

From philosophical perspectives, supervised learning is very similar to the case of student learning for which it is necessary to provide students with both questions and answers, such that students can identify if they have grasped the knowledge correctly, prior to exams. In this context, training data is like revision questions and test data is like exam questions.

Although supervised learning leads to straightforward evaluation of performance in a learning task, this type of learning would be constrained greatly in the big data era due to some issues, such as the rapid accumulation of varied data. More details on these issues will be given in Sect. 2.7.

H. Liu and M. Cocea, *Granular Computing Based Machine Learning*,
Studies in Big Data 35, https://doi.org/10.1007/978-3-319-70058-8_2

2.2   Heuristic Learning

Heuristic learning means learning with a specific strategy, which is based on statistical heuristics, such as Bayes Theorem (Eq. 2.1) and distance functions (Eqs. 2.2 and 2.3).

$$P(Y|X) = \frac{P(Y)P(X|Y)}{P(X)} \tag{2.1}$$

where X and Y are two events:

- $P(X)$ is read as the probability that event X occurs to be used as evidence supporting event Y.
- $P(Y)$ is read as the prior probability that event Y occurs on its own.
- $P(Y|X)$ is read as the posterior probability that event Y occurs given that event X truly occurs.
- $P(X|Y)$ is read as conditional probability that event X occurs subject to that event Y must occur.

$$D(x, y) = \sqrt{\sum_{i=1}^{n}(x_i - y_i)^2} \tag{2.2}$$

where n is the dimensionality (the number of attributes) and i is the index of the attribute. For example, if the data is in two dimensions, then Eq. 2.2 can be rewritten as Eq. 2.3 as follows:

$$D(x, y) = \sqrt{(x_1 - y_i)^2 + (x_2 - y_2)^2} \tag{2.3}$$

Bayes' Theorem is essentially used in Bayesian learning, and a popular algorithm of this type of learning is Naive Bayes (NB), as mentioned in Chap. 1. The learning outcome of this algorithm is finding the class that has the highest posterior probability given all input attributes with their values as the joint condition. The learning strategy of this algorithm is in the following procedure based on Bayes' theorem:

Step 1: Calculating the posterior probability of each class given each attribute with its value $P(y = c_k|x_i = v_{ij})$ on the basis of training instances, where y is the class attribute, c_k is the kth class label of y, x is the input attribute, i is the index of attribute x and v_{ij} is the jth value of x_i.
Step 2: Calculating the posterior probability of a class: $\prod_{i=0}^{n} P(y = c|x_i = v)$, where i is the index of attribute x and n is the number of attributes.
Step 3: Assigning the test instance the class that has the highest posterior probability on the basis of all the attribute values.

An example shown in Table 2.1 is used for illustrating the procedure of the NB algorithm:

Table 2.1 Illustrative Example for Naive Bayes [1]

x_1	x_2	x_3	y
0	0	1	0
0	1	0	1
1	0	1	1
1	1	0	0
0	1	1	1

Following Step 1 of the procedure illustrated above, the posterior probability of each class given each attribute with its value is listed below:

$P(y = 0|x_1 = 0) = 0.33,\ P(y = 1|x_1 = 0) = 0.67,$
$P(y = 0|x_1 = 1) = 0.50,\ P(y = 1|x_1 = 1) = 0.50,$
$P(y = 0|x_2 = 0) = 0.50,\ P(y = 1|x_2 = 0) = 0.50,$
$P(y = 0|x_2 = 1) = 0.33,\ P(y = 1|x_2 = 1) = 0.67,$
$P(y = 0|x_3 = 0) = 0.50,\ P(y = 1|x_3 = 0) = 0.50,$
$P(y = 0|x_3 = 1) = 0.33,\ P(y = 1|x_3 = 1) = 0.67,$

Following Step 2 of the above procedure, the value of y, when $x_1 = 1$, $x_2 = 1$ and $x_3 = 1$, is essentially calculated in the following way:

$P(y = 0|x_1 = 1, x_2 = 1, x_3 = 1) = P(y = 0|x_1 = 1) \times P(y = 0|x_2 = 1) \times P(y = 0|x_3 = 1) = 0.5 \times 0.33 \times 0.33;$
$P(y = 1|x_1 = 1, x_2 = 1, x_3 = 1) = P(y = 1|x_1 = 1) \times P(y = 1|x_2 = 1) \times P(y = 1|x_3 = 1) = 0.5 \times 0.67 \times 0.67;$

Following Step 3 of the above procedure, the following implication is made:

$P(y = 1|x_1 = 1, x_2 = 1, x_3 = 1) > P(y = 0|x_1 = 1, x_2 = 1, x_3 = 1) \rightarrow y = 1;$

Therefore, the test instance is assigned 1 as the value of y.

Distance measures have been popularly used in the K Nearest Neighbour (KNN) algorithm. The learning outcome of this algorithm is to assign the test instance the class which is the most commonly occurring from the k instances chosen from the training set. The learning strategy of this algorithm is in the following procedure:

Step 1: Choosing a method of measuring the distance between two points, e.g. Euclidean Distance.

Step 2: Determining the value of K, i.e. the number of training instances being selected.

Step 3: Finding the k instances (data points) that are closest to the given test instance.

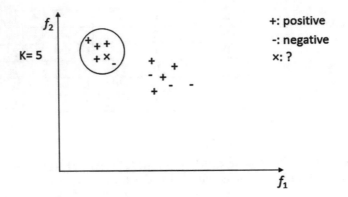

Fig. 2.1 K Nearest Neighbour for two class classification [1]

Step 4: Identifying the majority class that is most commonly occurring from the k instances chosen at step 2.

Step 5: Assigning the test instance the majority class identified at step 3.

An example is shown in Fig. 2.1 for illustrating the procedure of the KNN algorithm.

In the above example, the chosen value of K is 5 and there are two possible classes (positive or negative) towards classifying the test instance. As shown in Fig. 2.1, the five instances, which are closest to the test instance in terms of Euclidean Distance, are surrounded within a circle. In particular, there are four positive instances and one negative instance among the chosen ones. Therefore, the positive class is finally selected through majority voting towards classifying the above test instance.

On the basis of the above description, both NB and KNN would lead to deterministic classification. In other words, the repetitive conduct of the same experiment does not lead to variance of classification results, while both the training and test sets are fixed. The phenomenon of deterministic classification results indicates that heuristic learning generally leads to low variance but could result in high bias. More detailed analysis of bias and variance will be provided in Sect. 2.6, and some issues of heuristic learning in the big data era will be identified in Sect. 2.7.

2.3 Single-Task Learning

Single-task learning generally means that the learning outcome is single. In other words, the learning is aimed at predicting the value of a particular output attribute. In this context, there must be only a single test set involved in the testing stage. Also, both training and test sets can only contain one output attribute.

In traditional machine learning, classification is typically undertaken as a single task, which is simply aimed at classifying an instance into one of two or more categories. In practice, this kind of classification has been involved in broad application areas, such as medical diagnosis [2], pattern recognition [3], sentiment analysis [4–6] and ontology engineering [7].

On the other hand, traditional learning approaches can be categorized into single learning and ensemble learning. The former type involves learning of a single classifier whereas the latter type involves learning of an ensemble classifier. However, both types of classifiers are typically used for predicting the value of a single output attribute, so both single learning and ensemble learning would be considered to belong to single-task learning, unless there are multiple data sets involved in the testing stage or the data set being dealt with contains more than one output attribute.

In the era of big data, due to the increase in the complexity of data characteristics, single-task learning would be much limited against effective processing of big data, due to some issues. More details on these issues will be described in Sect. 2.7.

2.4 Discriminative Learning

Discriminative learning generally means learning to discriminate one class from other classes towards uniquely classifying an unseen instance. In other words, discriminative learning approaches work based on the assumption that different classes are mutually exclusive. In this context, each instance is considered to be clear-cut and thus can only belong to one class. Also, it is typical to force all instances to be classified, i.e. it is not acceptable to leave an instance unclassified.

As identified in [8], the assumption of mutual exclusion among different classes does not always hold in real applications. For example, a movie on war can belong to both military and history, due to the fact that this type of movies tells a real story that involves soldiers and that happened in the past. Also, the same book may be used by students from different departments and thus can belong to different subjects.

On the other hand, while different classes are truly mutually exclusive, instances may be complex leading to difficulty in classifying each of such instances to one category only. For example, in handwritten digits recognition, handwritten '4' and '9' are very similar and sometimes hard to distinguish. From this point of view, it is not appropriate to assume each instance to be clear-cut.

Moreover, a real data set is dynamically grown in practice. In this context, if a classifier is trained using a data set with a certain number of predefined classes, then it is likely to result in inaccurate classifications when the data set is updated [9, 10]. In fact, it is very likely to occur in practice that a data set is initially assigned a number of class labels by domain experts on the basis of their incomplete knowledge, but the data set is later provided with extra labels following the gain of new knowledge.

In the above case, it is not appropriate to force all instances to be classified since it is possible that an instance does not belong to any one of the initially assigned classes. The above case also indicates that the learning of a classifier must be redone once the class labels of the data set are updated, i.e. the classifier is poor in maintainability, reusability, extendability and flexibility and thus not acceptable in real applications [8].

In order to address the above issues, multi-label classification has recently been proposed. The term multi-label generally means that an instance is assigned multiple class labels jointly or separately, which is the main difference to the term single-label meaning that an instance is assigned a single class label only. Details can be found in [11–14].

There are three typical ways of dealing with multi-label classification problems referred to as PT3, PT4 and PT5 respectively, as reviewed in [11].

PT3 is designed to enable that a class consists of multiple labels as illustrated in Table 2.2. For example, two classes A and B can make up three labels: A, B and $A \wedge B$. PT4 is designed to do the labelling on the same data set separately regarding each of the predefined labels as illustrated in Tables 2.3 and 2.4.

In addition, PT5 is aimed at uncertainty handling. In other words, it is not certain to which class label an instance should belong, so the instance is assigned all the possible labels and is treated as several different instances that have the same inputs but different class labels assigned. An illustrative example is given in Table 2.5: both instances (3 and 4) appear twice with two different labels (A and B) respectively, which would be treated as four different instances (two assigned A and the other two assigned B) in the process of learning.

In the big data era, all of the above approaches for mutli-label classification would lead to negative impacts on learning of classifiers. More details on these impacts will be given in Sect. 2.7.

Table 2.2 Example of PT3 [8]

Instance ID	Class
1	A
2	B
3	$A \wedge B$
4	$A \wedge B$

Table 2.3 Example of PT4 on Label A [8]

Instance ID	Class
1	A
2	$\neg A$
3	A
4	A

Table 2.4 Example of PT4 on Label B [8]

Instance ID	Class
1	$\neg B$
2	B
3	B
4	B

Table 2.5 Example of PT5 [8]

Instance ID	Class
1	A
2	B
3	A
3	B
4	A
4	B

2.5   Random Data Partitioning

In machine learning, there are several ways of data partitioning for experimentation. The most popular ways are typically referred to as training/test partitioning or cross-validation [15–17].

The training/test partitioning approach typically involves the partitioning of data into a training set and a test set in a specific ratio, e.g. 70% of the data is used as the training set and 30% of the data is used as the test set. This data partitioning can be done randomly or in a fixed way (e.g. the first 70% of the instances in the dataset are assigned to training set and the rest to the test set). The fixed way is typically avoided (except when order matters) as it may introduce systematic differences between the training set and the test set, which leads to sample representativeness issues. To avoid such systematic differences, the random assignment of instances into training and test sets is typically used.

As mentioned in Chap. 1, cross-validation is conducted by partitioning a data set into n folds (or subsets), followed by an iterative process of combining the folds into different training and test sets. Due to the case that cross-validation is generally more expensive in terms of computational cost, the training/test partitioning approach is often used instead. Also, there have been some new perspectives of using the above two ways of data partitioning identified in [18]. In other words, cross-validation and the training/test partitioning approach would be used for different purposes in machine learning experimentation, and more details on the differences will be presented and discussed in Chap. 8.

In the big data era, random partitioning of data could lead to more serious issues that affect the performance of learning algorithms. More details on these issues will be given in Sect. 2.7.

2.6   General Issues

In traditional machine learning, prediction errors can be caused by bias and variance. Bias generally means errors originated from heuristics of learning algorithms and variance means errors originated from random processing of data. In terms of learning performance, heuristic learning introduced in Sect. 2.2 generally leads to high bias but low variance, and random partitioning of data introduced in Sect. 2.5 generally leads to low bias but high variance. From this point of view, it has become necessary to manage the trade-off between bias and variance, towards optimizing the performance of learning.

A principal problem of machine learning is referred to as overfitting [19], which essentially means that a model performs to a high level of accuracy on training data but to a low level of accuracy on test data. It was claimed traditionally [20] that the cause of overfitting would be due to the bias originated from the heuristics of the employed learning algorithm leading to the poor generality of the learned model. However, we argue that the overfitting problem is also caused due to the sample representativeness issue, i.e. a huge dissimilarity between training and test sets.

For example, as illustrated in Fig. 2.2, the hypothesis space is covered by the whole data set, and the training space results from the distribution of the training instances, which is a small subspace of the hypothesis space.

In this case, if the test instances are all distributed inside the training space, which means that the test instances are highly similar to the training instances, then it would be very likely that the model would perform to a high level of classification accuracy. However, if most test instances are located far away from the training space, which means that the test instances are mostly dissimilar to these training instances, then it would be very unlikely for the model to perform to a high level of classification accuracy.

In Sect. 2.7, we will analyze further the impact of the sample representativeness issue on learning performance in the big data era.

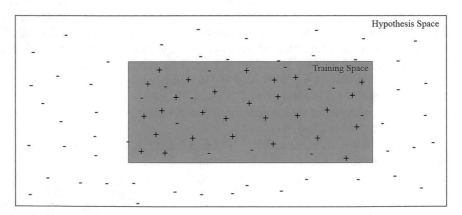

Fig. 2.2 Illustration of Overfitting (NB: "+" indicates training instance and "-" indicates test instance)

2.7 Impacts from Big Data

In the big data era, learning tasks have become increasingly more complex. In this context, traditional machine learning has been too shallow to deal with big data sufficiently. In this section, we analyze how the traditional learning approaches presented in Sects. 2.1–2.5 would be constrained in the context of big data processing.

In terms of supervised learning, as mentioned in Sect. 2.1, the data used in a learning task must be fully labelled by experts. However, due to the vast and rapid increase in the size of data, it has become an overwhelming task to have all instances labelled by experts. From this point of view, it has been necessary to turn supervised learning into semi-supervised learning, towards involving machine based labelling instead of using only expert based labelling. In other words, a small part of a large data set is labelled by experts, and the rest of the data set is labelled by using learning algorithms on the basis of the experts labelled data subset. In Chap. 3, we will present how to achieve semi-supervised Learning through machine based labelling.

In terms of heuristic learning, as mentioned in Sect. 2.2, it could lead to high bias and low variance in terms of learning performance, when both training and test sets are fixed. However, in the big data era, the characteristics of data could be changed quickly and significantly. In this context, high bias would lead to the case that a learning approach is highly suitable for dealing with the original data set but becomes very unsuitable later due to the significant change of data characteristics (high variability of data). From this point of view, it has become necessary to turn heuristic learning into semi-heuristic learning towards achieving the trade-off between bias and variance. In Chap. 4, we will present how to achieve semi-heuristic learning through nature inspired techniques. Also, we will show in Chap. 8 how greatly the high variability of data can impact on the performance of heuristic learning approaches.

In terms of single-task learning, as mentioned in Sect. 2.3, both training and test sets can only have one output attribute.

In the big data era, classification can be from different perspectives, which indicates that it has become normal to learn from the same feature set for different classification tasks. For example, in the context of student classification, we can classify a student in terms of their study mode (full-time or part-time), their learning performance (good or bad) and their degree level (undergraduate or postgraduate), on the basis of the same feature set. From this point of view, it has become necessary to turn single-task learning into multi-task learning.

Also, single task learning requires that there must be only a single test set involved in the testing stage. In the context of big data processing, due to a large size of data, it has been necessary to decompose a data set into different subsets [10], such that learning from different subsets can be done in parallel. From this point of view, each subset is divided into a training subset and a test subset for a particular learning task, which again indicates the necessity of turning single-task learning into multi-task learning.

In Chap. 5, we will present how to achieve multi-task learning through using fuzzy approaches.

In terms of discriminative learning, as mentioned in Sect. 2.4, it is based on the assumption that different classes are mutually exclusive. Since the assumption does not always hold for practical applications, multi-label learning approaches (PT3, PT4 and PT5) have been proposed for addressing this issue. As argued in [8], these approaches are still aimed at dealing with classification problems in a discriminative way, i.e. the aim is still to learn a classifier that provides a unique output, in terms of the class to which an instance belongs.

In the big data era, PT3 may result in a massive number of classes, i.e. $2^n - 1$, where n is the number of class labels. PT4 may result in class imbalance. For example, a balanced data set involves three classes A, B and C, and thus the frequency ($\frac{1}{3}$) of class A is far lower than the one ($\frac{2}{3}$) of class ¬A (i.e. B+C). PT5 may result in a massive size of training sample leading to high computational complexity. From software engineering perspectives, PT3 may result in high coupling, while different class labels that are not correlated are merged into a new class. Coupling generally refers to the degree of interdependence between different parts [21]. Similarly, PT4 may result in low cohesion, while different class labels that are correlated get separated. Low cohesion means the degree to which the parts of a whole link together is lower [21], and thus failing to identify the correlations between different classes.

On the basis of the above argumentation on discriminative learning, it has been necessary to address further the issue of mutual exclusion among classes. In particular, we will present in Chap. 5 how to adopt generative learning through fuzzy approaches towards dealing with the above issue.

In terms of random data partitioning, as mentioned in Sect. 2.5, it is typically conducted by selecting training/test instances in a particular probability, e.g. if the distribution between training and test sets needs to be 70/30, then an instance has a 70% chance to be selected as a training one. In the big data era, this way of partitioning is more likely to lead to two major issues: (a) class imbalance and (b) sample representativeness issues.

The first issue of class imbalance [22, 23] is known to affect the performance of many learning algorithms. Randomly partitioning the data, however, can lead to class imbalance in both the training set and the test set, even when there is no imbalance in the overall data set. For example, let us consider a 2-class (e.g. positive and negative class) data set with a balanced distribution of instances across classes, i.e. 50% of the instances belong to the positive class and 50% of the instances belong to the negative class. When the data set is partitioned by selecting training/test instances randomly, it is likely that the class balance of the data set will be broken, which would lead, for example, to more than 50% of the training instances belonging to the positive class and more than 50% of the test instances belonging to the negative class, i.e. the training set has more positive instances than negative ones, while the test set has the opposite situation.

The second issue is about sample representativeness and the fact that the random partitioning may lead to high dissimilarity between training and test instances, as mentioned in Sect. 2.6. In the context of student learning, the training instances are like the revision questions and the test instances are like the exam questions. To test effectively the performance of student learning, the revision questions should be representative with respect to the learning content covered in the exam questions. The random partitioning of data, however, can result in the case that the training instances are dissimilar to the test instances, which corresponds to the situation that students are tested on what they have not yet learned. Such a situation not only leads to a poor performance, but also to a poor judgment of the learner capability. Thus, in the context of predictive modelling, some algorithms may be judged as not being suitable for a particular problem due to a poor performance, when in reality the poor results are not due to the algorithm, but to the representativeness of the training sample.

In order to address the above two issues, it has been necessary to turn random partitioning into semi-random partitioning. In Chap. 6, we will present a multi-granularity framework for semi-random partitioning of data, which focuses on dealing with the class imbalance issue and provides a brief proposal towards dealing effectively with the sample representativeness issue.

References

1. H. Liu, A. Gegov, and M. Cocea. 2016. Nature and biology inspired approach of classification towards reduction of bias in machine learning. In *International Conference on Machine Learning and Cybernetics*, Jeju Island, South Korea, 10-13 July 2016, pp. 588–593.
2. Cendrowska, J. 1987. Prism: an algorithm for inducing modular rules. *International Journal of Man-Machine Studies* 27: 349–370.
3. Z. Teng, F. Ren, and S. Kuroiwa. 2007. Emotion recognition from text based on the rough set theory and the support vector machines. In *International Conference on Natural Language Processing and Knowledge Engineering*, Beijing, China, 30 August- 1 September 2007, pp. 36–41.
4. K. Reynolds, A. Kontostathis, and L. Edwards. 2011. Using machine learning to detect cyberbullying. In *Proceedings of the 10th International Conference on Machine Learning and Applications*, December 2011, pp. 241–244.
5. Tripathy, A., A. Agrawal, and S.K. Rath. 2015. Classication of sentimental reviews using machine learning techniques. *Procedia Computer Science* 57: 821–829.
6. H. Liu and M. Cocea. 2017. Fuzzy rule based systems for interpretable sentiment analysis. In *International Conference on Advanced Computational Intelligence*, Doha, Qatar, 4–6 February 2017, pp. 129–136.
7. Pedrycz, W., and S.M. Chen. 2016. *Sentiment Analysis and Ontology Engineering: An Environment of Computational Intelligence*. Heidelberg: Springer.
8. H. Liu, M. Cocea, A. Mohasseb, and M. Bader. 2017. Transformation of discriminative single-task classification into generative multi-task classification in machine learning context. In *International Conference on Advanced Computational Intelligence*, Doha, Qatar, 4–6 February 2017, pp. 66–73.
9. Wang, X., and Y. He. 2016. Learning from uncertainty for big data: Future analytical challenges and strategies. *IEEE Systems, Man and Cybernetics Magazine* 2 (2): 26–32.
10. Suthaharan, S. 2014. Big data classification: problems and challenges in network intrusion prediction with machine learning. *ACM SIGMETRICS Performance Evaluation Review* 41 (4): 70–73.
11. Tsoumakas, G., and I. Katakis. 2007. Multi-label classification: an overview. *International Journal of Data Warehousing and Mining* 3 (3): 1–13.

12. I. K. G. Tsoumakas and I. Vlahavas. 2010. Mining multi-label data. In *Data Mining and Knowledge Discovery Handbook*. Springer 2010, pp. 667–685.
13. Boutell, M.R., X.S.J. Luo, and C.M. Brown. 2004. Learning multi-label scene classification. *Pattern Recognition* 37: 1757–1771.
14. Zhang, M., and H. Zhou. 2014. A review on multi-label learning algorithms. *IEEE Transactions on Knowledge and Data Engineering* 26 (8): 1819–1837.
15. R. Kohavi. 1995. A study of cross-validation and bootstrap for accuracy estimation and model selection. In *Proceedings of the Fourteenth International Joint Conference on Artificial Intelligence*. San Mateo, CA: Morgan Kaufmann, pp. 1137–1143.
16. Geisser, S. 1993. *Predictive Inference*. New York: Chapman and Hall.
17. Devijver, P.A. 1982. *Pattern Recognition: A Statistical Approach*. London: Prentice Hall.
18. Liu, H., A. Gegov, and M. Cocea. 2017. Unified framework for control of machine learning tasks towards effective and efficient processing of big data. In *Data Science and Big Data: An Environment of Computational Intelligence*, pp. 123–140. Switzerland: Springer.
19. Liu, H., A. Gegov, and M. Cocea. 2016. *Rule Based Systems for Big Data: A Machine Learning Approach*. Switzerland: Springer.
20. Mitchell, T. 1997. *Machine Learning*. New York: McGraw Hill.
21. T. C. Lethbridge and R. Laganire. 2005. *Object Oriented Software Engineering: Practical Software Development using UML and Java (2nd)*. Maidenhead: McGraw-Hill Education.
22. Longadge, R., S.S. Dongre, and L. Malik. 2013. Class imbalance problem in data mining: Review. *International Journal of Computer Science and Network* 2 (1): 83–87.
23. Ali, A., S.M. Shamsuddin, and A.L. Ralescu. 2015. Classification with class imbalance problem: a review. *International Journal of Advanced Soft Computing Applications* 7 (3): 176–204.

Chapter 3
Semi-supervised Learning Through Machine Based Labelling

Abstract In this chapter, we describe the concepts of semi-supervised learning and show the motivation of developing semi-supervised learning approaches in the context of big data. We also review existing approaches of semi-supervised learning and then focus the strategy of semi-supervised learning on machine based labelling. Furthermore, we present two proposed frameworks of semi-supervised learning in the setting of granular computing, and discuss the advantages of the frameworks.

3.1 Overview of Semi-supervised Learning

Semi-supervised learning generally means that the training data is not fully labelled, i.e. it is a combination of labelled instances and unlabelled instances. In the big data era, semi-supervised learning would be increasingly more needed, since it is overwhelming to have all training instances labelled by experts, due to a huge size of data, as argued in Chap. 2.

As introduced in [1], semi-supervised learning took off in 1970s, due to the consideration of the problem that the Fisher linear discriminant rule needs to be estimated with use of unlabelled data. Moreover, semi-supervised learning became more interesting to researchers in 1990s, mostly due to practical issues in natural language processing and text classification.

As organized in [1], semi-supervised learning methods can be divided into four classes, namely, self-training, generative models, S3VM, graph-based algorithms and multi-view algorithms.

Self-training is also referred to as self-learning, which typically involves four steps [1] as follows:

1. Choose a supervised learning method and learn a classifier f from a labelled set of training instances (x_l, y_l).
2. Classify all unlabelled instances $x \in X_u$, by using f.
3. Select an instance x^* with the highest confidence, and add the instance with its label $(x^*, f(x^*))$ to the labelled set of training instances.
4. Repeat Step 2–3.

© Springer International Publishing AG 2018
H. Liu and M. Cocea, *Granular Computing Based Machine Learning*,
Studies in Big Data 35, https://doi.org/10.1007/978-3-319-70058-8_3

In application areas, self-learning has been used for object detection [2], hierarchi-cal multi-label classification [3], training decision tree classifiers [4], Classification of Protein Crystallization Imagery [5], nearest neighbour classification [6] and sen-timent classification [7]. More details on self-learning can be found at [8].

Since self-learning is essentially aimed at labelling training instances, through using a classifier built by using a supervised learning algorithm, we refer to this way of labelling as machine based labelling in the rest of this book.

Generative models work for semi-supervised learning based on the assumption that each class has a Gaussian distribution in feature space [1], towards identifying the most likely decision boundary. Some examples of generative models include: (a) mixture of Gaussian distributions (based on Gaussian mixture model), which has been popularly used for image classification [9], (b) mixture of multinomial distributions (based on Naive Bayes), which has been popularly used for text cate-gorization [10], and (c) hidden markov models, which has been popularly used for speech recognition [11].

In addition, S3VM works for semi-supervised learning based on the assumption that unlabelled instances of different classes are separated with a large margin [1]. Graph-based algorithms work for semi-supevised learning based on the assumption that a graph is drawn on the basis of both labelled and unlabelled data, and instances that are connected by heavy edges tend to be assigned the same label [1]. This type of semi-supervised learning methods has been used in both text classification [12] and handwritten digits recognition [13].

Multi-view algorithms typically involve co-training, which means that an instance can be represented by two different sets of features [1]. In this context, a classifier is learned from each set of features, and these classifiers then involve further mutual learning (learn from each other). As introduced in [1], multi-view algorithms work for semi-supervised learning based on the assumption that each of these different feature sets is of good quality leading to learning of a good classifier, and that these feature sets are conditionally independent in terms of their relationship to the class attribute.

In this chapter, we focus on machine based labelling for semi-supervised learning. In particular, we will present two approaches of machine based labelling in Sect. 3.2 and discuss the key features of these approaches in Sect. 3.3.

3.2   Granular Framework of Learning

As mentioned in Sect. 3.1, machine based labelling is a main approach to achieve semi-supervised learning towards dealing effectively with unlabelled training instances. In this section, we present our proposed framework of semi-supervised learning through machine based labelling, and two specific approaches based on K Nearest Neighbour (KNN) and Support Vector Machine (SVM), respectively.

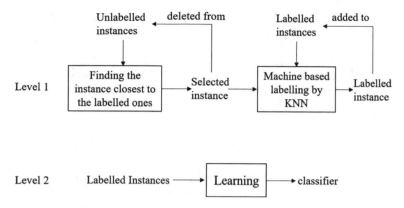

Fig. 3.1 Semi-supervised Learning through KNN

In general, the proposed semi-supervised learning framework involves two levels of granularity as follows:

1. Level 1: machine based labelling through using a supervised learning approach for dealing with unlabelled training instances
2. Level 2: learning models from both expert labelled instances and machine labelled instances

In the setting of granular computing, granulation is to divide a training set into a subset of labelled instances and a subset of unlabelled instances. Organization is to merge a set of expert labelled instances and a set of machine labelled instances into a whole training set.

In terms of machine based labelling, KNN and SVM would be very good fits towards dealing with unlabelled instances. In particular, the procedure of the KNN based approach of machine based labelling involves the following steps:

Step 1: Find and select the unlabelled instance that is closest to the labelled instances on average.

Step 2: Learn a classifier from the labelled instances.

Step 3: Classify the unlabelled instance selected at Step 1 by using the classifier learned at Step 2.

Step 4: Add the machine labelled instance into the labelled training set, and delete the instance from the unlabelled training set.

Step 5: Repeat Step 1–4 until no unlabelled instances remain.

The corresponding double granularity framework of KNN based semi-supervised learning is illustrated in Fig. 3.1

Also, the procedure of the SVM based approach of semi-supervised learning involves the following steps:

Step 1: Learn a classifier from the labelled instances by identifying an optimal boundary for separating instances of different classes.

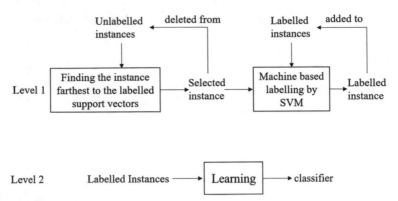

Fig. 3.2 Semi-supervised Learning through SVM

Step 2: Find and select the unlabelled instance that is farthest to the classification boundary identified at Step 1.

Step 3: Classify the unlabelled instance selected at Step 2 by using the classifier learned at Step 1.

Step 4: Add the machine labelled instance into the labelled training set, and delete the instance from the unlabelled training set.

Step 5: Repeat Step 1–4 until no unlabelled instances remain.

The corresponding double granularity framework of SVM based semi-supervised learning is illustrated in Fig. 3.2.

3.3 Discussion

In the context of machine based labelling, the most critical thing is to minimize the risk to assign incorrect classes to unlabelled instances, i.e. to avoid learning from incorrectly labelled instances. From this point of view, we argue that it is necessary to involve incremental learning for classifying unlabelled instances one by one. In other words, at each iteration, only one unlabelled instance is selected, and a classifier is learned from the current set of labelled instances towards classifying the unlabelled instance. In this context, it is expected that the selected instance is the one most likely to be classified correctly towards minimizing the risk of incorrect labelling of the subsequent unlabelled instances.

In order to avoid incorrect labelling of instances, we argue that it is necessary the method employed for machine based labelling only needs to learn a classifier from a few instances, due to the case that there is only a small number of labelled instances. From this point of view, lazy learning methods, such as KNN and SVM, are more suitable for machine based labelling, since both methods only need to examine a few training instances towards learning a classifier, i.e. KNN only needs to find k

instances that are closest to the unseen instance being classified, and SVM only needs to select a few instances as support vectors towards learning a classifier.

In terms of KNN based semi-supervised learning, in order to minimize the risk of incorrect labelling of instances, we design to select and classify at each iteration the unlabelled instance that is closest to the labelled instances on average, i.e. the unlabelled instance, which is closest to the labelled instances on average, is considered to be the one most likely to classified correctly.

In terms of SVM based semi-supervised learning, in order to minimize the risk of incorrect labelling of instances, we design to select and classify at each iteration the unlabelled instance that is farthest to the support vectors on average (farthest to the classification boundary learned by using SVM), i.e. the unlabelled instance, which is farthest to the support vectors on average, is considered to be the one most unlikely to classified incorrectly.

Overall, machine based labelling involves a chained impact leading to incorrect labelling of instances, if any one of the unlabelled instances is assigned a wrong class. In other words, the case that one instance is labelled incorrectly would usually increase the chance of incorrect labelling of the subsequent instances. In order to avoid the above case, it is necessary to involve incremental learning towards labelling instances one by one. In this way, at each iteration, it is expected to minimize the risk of incorrect labelling of the selected instance, such that the overall correctness of machine based labelling would be optimized, i.e. the number of correctly labelled instances is maximized.

References

1. Zhu, X., and A.B. Goldberg. 2009. *Introduction to semi-supervised learning*. San Rafael: Morgan and Claypool Publishers.
2. Rosenberg, C., M. Hebert, and H. Schneiderman. 2005. Semi-supervised self-training of object detection models. In IEEE Workshop on Applications of Computer Vision. *Breckenridge, CO, USA 5-7*: 29–36.
3. A. M. Santos and A. M. P. Canuto. 2014. Applying the self-training semi-supervised learning in hierarchical multi-label methods. In *IEEE International Joint Conference on Neural Networks*, Beijing, China, 6-11 July 2014, 873–879.
4. Tanha, J., M. van Someren, and H. Afsarmanesh. 2017. Semi-supervised self-training for decision tree classifiers. *International Journal of Machine Learning and Cybernetics* 8 (1): 355–370.
5. M. Sigdel, I. Dinç, S. Dinç, M. S. Sigdel, M. L. Pusey, and R. S. Aygün. 2014. Evaluation of semi-supervised learning for classification of protein crystallization imagery. In *Proceedings of IEEE Southeastcon*, 1–6.
6. I. Triguero, J. A. Sez, J. Luengo, S. Garca, and F. Herrera. 2014. On the characterization of noise filters for self-training semi-supervised in nearest neighbor classification. *Neurocomputing*, 132 (1):30–41.
7. Z. Liu, X. Dong, Y. Guan, and J. Yang. 2013. Reserved self-training: A semi-supervised sentiment classification method for chinese microblogs. In *International Joint Conference on Natural Language Processing*, Nagoya, Japan, 14-18 October 2013, 455–462.
8. Prakash, V.J., and L. Nithya. 2014. A survey on semi-supervised learning techniques. *International Journal of Computer Trends and Technology* 8 (1): 25–29.

9. Cho, W., S. Seo, I. Na, and S. Kang. 2014. *Automatic images classification using HDP-GMM and local image features*. *In Ubiquitous Information Technologies and Applications*, 323–333. Heidelberg: Springer.
10. A. M. Kibriya, E. Frank, B. Pfahringer, and G. Holmes. 2004. Multinomial naive bayes for text categorization revisited. In *Australasian Joint Conference on Artificial Intelligence*, Cairns, Australia, 4-6 December 2004, 488–499.
11. Rabiner, L.R., and B.H. Juang. 1992. Hidden markov models for speech recognition: Strengths and limitations. In *Speech Recognition and Understanding*, 3–29. Heidelberg: Springer.
12. K. S. Thakkar, R. Dharaskar, and M. Chandak. 2010. Graph-based algorithms for text summarization. In *Third International Conference on Emerging Trends in Engineering and Technology*, Goa, India, 19–21 November 2010, 516–519.
13. A. Filatov, A. Gitis, and I. Kil. 1995. Graph-based handwritten digit string recognition. In *Third International Conference on Document Analysis and Recognition*, Montreal, Que., Canada, 14–16 August 1995, 845–848.

Chapter 4
Nature Inspired Semi-heuristic Learning

Abstract In this chapter, we describe the concepts of nature inspired semi-heuristic learning by using voting based learning methods as examples. We also present a nature inspired framework of ensemble learning, and discuss the advantages that nature inspiration can bring into a learning framework, from granular computing perspectives.

4.1 Overview of Semi-heuristic Learning

Semi-heuristic learning generally means that a learning approach is partially based on heuristics and partially random. In the context of nature inspired semi-heuristic learning, the methods typically involve voting for classifying instances, and employ probabilistic voting through nature inspiration, towards reduction of the bias in machine learning.

As mentioned in Chap. 2, Bagging and Boosting employ majority voting and weighted voting, respectively. Both voting methods can be seen as deterministic voting, since they both work in the context of deterministic logic by assuming that there is no uncertainty for classifying an unseen instance.

Probabilistic voting [1] is considered to be inspired by nature and biology in the context of granular computing, since the voting is made on the basis of the hypothesis that the class with the highest frequency or weight only has the best chance of being selected towards classifying an unseen instance. In other words, it is not guaranteed that the class with the highest frequency or weight will definitely be selected to be assigned to the unseen instance.

In this chapter, we present the use of probabilistic voting for both Bagging and Boosting towards improving the overall classification accuracy. In particular, majority voting (involved in Bagging) and weighted voting (involved in Boosting) are both replaced with probabilistic voting. The procedure of probabilistic voting is illustrated below:

Step 1: calculating the weight W_i for each single class i.
Step 2: calculating the total weight W over all classes.

© Springer International Publishing AG 2018
H. Liu and M. Cocea, *Granular Computing Based Machine Learning*,
Studies in Big Data 35, https://doi.org/10.1007/978-3-319-70058-8_4

Step 3: calculating the percentage of weight P_i for each single class i, i.e. $P_i = W_i \div W$.

Step 4: Randomly selecting a single class i with the probability P_i towards classifying an unseen instance.

The following example relating to Bayes Theorem is used for the illustration of the above procedure:

Inputs(binary): x_1, x_2, x_3
Output(binary): y

Probabilistic Correlation:
$P(y = 0|x_1 = 0) = 0.4, P(y = 1|x_1 = 0) = 0.6, P(y = 0|x_1 = 1) = 0.5,$
$P(y = 1|x_1 = 1) = 0.5;$
$P(y = 0|x_2 = 0) = 0.7, P(y = 1|x_2 = 0) = 0.3, P(y = 0|x_2 = 1) = 0.6,$
$P(y = 1|x_2 = 1) = 0.4;$
$P(y = 0|x_3 = 0) = 0.5, P(y = 1|x_3 = 0) = 0.5, P(y = 0|x_3 = 1) = 0.8,$
$P(y = 1|x_3 = 1) = 0.2;$

While $x_1 = 0, x_2 = 1, x_3 = 1, y =?$
Following Step 1, the weight W_i for each single value of y is:
$P(y = 0|x_1 = 0, x_2 = 1, x_3 = 1) = P(y = 0|x_1 = 0) \times P(y = 0|x_2 = 1) \times$
$P(y = 0|x_3 = 1) = 0.4 \times 0.6 \times 0.8 = 0.192$
$P(y = 1|x_1 = 0, x_2 = 1, x_3 = 1) = P(y = 1|x_1 = 0) \times P(y = 1|x_2 = 1) \times$
$P(y = 1|x_3 = 1) = 0.6 \times 0.4 \times 0.2 = 0.048$

Following Step 2, the total weight W is $0.24 = 0.192 + 0.048$.

Following Step 3, the percentage of weight P_i for each single value of y is:
Percentage for $y = 0$: $P_0 = 0.192 \div 0.24 = 80\%$
Percentage for $y = 1$: $P_1 = 0.048 \div 0.24 = 20\%$

Following Step 4, $y = 0$ (80% chance) or $y = 1$ (20% chance).

In the above illustration, both majority voting and weighted voting would result in 0 being assigned to y due to its higher frequency or weight shown in Step 4. In particular, in the context of majority voting, Step 4 would indicate that the frequency for y to equal 0 is 80% and the one for y to equal 1 is 20%. Also, in the context of weighted voting, Step 4 would indicate that over the total weight the percentage of the weight for y to equal 0 is 80% and the percentage of the weight for y to equal 1 is 20%. Therefore, both types of voting would choose to assign y the value of 0. However, in the context of probabilistic voting, Step 4 would indicate that y could be assigned either 0 (with 80% chance) or 1 (with 20% chance). In this way, the bias in voting can be effectively reduced towards improvement of overall accuracy of classification in ensemble learning.

The probabilistic voting approach illustrated above is very similar to natural selection which is one step of the procedure of genetic algorithms [2], i.e. the probability of a class being selected is very similar to the fitness of an individual involved in natural selection. In particular, the way of selecting a class involved in Step 4 of the above procedure is inspired by the Roulette Wheel Selection [3].

Probabilistic voting has been used in [1] for advancing individual learning algorithms that involve voting, such as Naive Bayes and K Nearest Neighbour, and the results are shown in Table 4.1 and Fig. 4.1. In particular, NB II and KNN II mean that Naive Bayes and K Nearest Neighbour employ probabilistic voting for classifying instances. In other words, majority voting is employed for NB I and weighted voting is employed for KNN I.

Table 4.1 shows that for both Naive Bayes and K Nearest Neighbour the probabilistic voting manages to increase the classification accuracy in comparison with the other two voting strategies in most cases. In particular, for Nave Bayes, the probabilistic voting manages effectively to achieve higher accuracy than the weighted voting does in 14 out of 15 cases, except for the case of credit-g data set for which the accuracy is the same when the two voting methods are compared. For K Nearest Neighbour, there are 12 out of 15 cases in which the probabilistic voting manages effectively to achieve higher accuracy than the majority voting and 2 cases where the two voting methods perform the same.

In addition, Fig. 4.1 shows that for K Nearest Neighbour the probabilistic voting usually manages to achieve higher accuracy than the majority voting does while the K value is changed incrementally from 2 to 5. It can also be seen in Fig. 4.1 that the probabilistic voting generally manages to keep a similar level of variance or even a lower level in comparison with the majority voting.

Table 4.1 Classification accuracy [1]

Dataset	NB I (%)	NB II (%)	KNN I (%)	KNN II (%)
anneal	88	92	95	97
audiology	72	74	59	65
autos	48	65	54	67
breast-cancer	67	73	70	75
breast-w	95	98	95	97
colic	71	80	76	80
credit-a	75	81	85	83
credit-g	75	75	73	73
ecoli	84	90	85	86
glass	42	60	63	70
heart-c	86	88	87	87
heart-h	86	87	78	85
heart-stalog	86	91	84	85
hepatitis	83	92	89	95
ionosphere	81	91	77	90

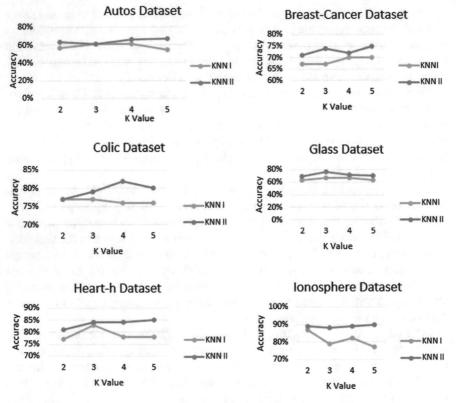

Fig. 4.1 Performance of KNN with different K values [1]

The motivation of proposing the probabilistic voting is to make the voting based classification more natural. In this way, bias originated from learning algorithms would be reduced effectively, leading to reduction of overfitting of training data, as indicated in Table 4.1 and Fig. 4.1. Since this type of voting is involved in ensemble learning as well, probabilistic voting could also lead to improved results in this context. More details on using probabilistic voting for advancing ensemble learning are presented in Sect. 4.2.

4.2   Granular Framework of Learning

A nature inspired framework of ensemble learning was proposed in [4]. In particular, probabilistic voting was applied to Bagging and Boosting by replacing majority voting and weighted voting, respectively, with probabilistic voting, for classifying unseen instances. The nature inspired Bagging approach is illustrated in Fig. 4.2.

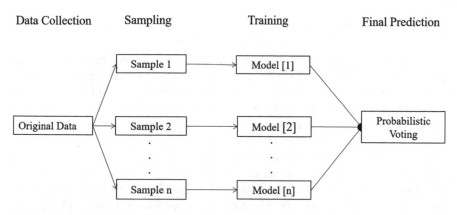

Fig. 4.2 Nature inspired bagging approach

In the setting of granular computing, the Bagging approach involves explicitly both granulation and organization. In particular, granulation is aimed at drawing a number of samples with replacement from training data, i.e. to get different versions of training data. Organization is aimed at combining the models learned from different samples into an ensemble model, towards classifying unseen instances through voting.

The Bagging framework involves two levels of granularity:

1. Level 1: From each sample of training data, a model is learned by the employed algorithm.
2. Level 2: The fitness of each model is learned for measuring its confidence of correctly classifying instances of a specific class.

In probabilistic voting, the concept of probabilistic intervals, which is a part of granular computing theory, is used. In particular, each class, which is viewed as a probabilistic granule, is assigned an interval based on its weight calculated in Step 1 of probabilistic voting, and the class would be selected and assigned to an unseen instance if a random number generated is within this interval.

For example, there are two classes (0 and 1). According to the illustration of probabilistic voting in Sect. 4.1, the chance for assigning 0 and 1 is 80% and 20%, respectively. In the context of probabilistic intervals, the interval [1, 20] is assigned to class '0' and the interval [21, 100] is assigned to class '1'. If a randomly generated integer is within [1, 20], then class '0' is selected and assigned to the unseen instance being classified. Otherwise, class '1' is selected and assigned to the unseen instance.

4.3 Discussion

In the context of granular computing, the frequency of a class can be viewed as an information granule that enables the class to be selected for being assigned to a test instance. Similarly, the weight of a class can be viewed as a part of information granules that enable the class to be selected towards classifying an unseen instance. From this point of view, the class with the highest frequency of being predicted by base classifiers means to have been assigned the most information granules that enable the class to be selected for being assigned to a test instance. Similarly, the class with the highest weight means to have been assigned the highest percentage of the information granules that enable the class to be selected towards classifying an unseen instance. More details on information granules can be found in [5–7].

As mentioned in Chap. 1, for classifying test instances, the Bagging method is biased to always select the most frequently occurring class and the Boosting method is biased to always select the most highly weighted class. This is due to the assumption that all the independent predictions by the base classifiers provide a complete and highly trusted set of information granules, each of which votes towards one class and against all the other classes. However, it is fairly difficult to guarantee that a set of granules is complete, due to the fact that the training and validation sets are very likely to be incomplete in practice. Also, it is commonly known that a training set may be imbalanced, due to insufficient collection of data, which is likely to result in a class being assigned much more information granules than the other classes. In addition, a learning algorithm may not be suitable to learn a model on a particular sample set. In this case, the information granules, which are provided from the predictions of the models learned by that algorithm, would be much less trusted.

In ensemble learning, Bagging needs to draw a number of samples of the original data on a random basis and Boosting needs to iteratively evaluate the weight of training instances. The nature of the Bagging method may result in poor samples of training data being drawn in terms of incompleteness and imbalance. The nature of the Boosting method may result in poor evaluation of training instances in terms of their weights. If the employed learning algorithms are not suitable to the sampled data for Bagging or the weighted data for Boosting, then the frequency or the weight of classes would be much less trusted for classifying test instances. Therefore, the majority voting involved in Bagging and the weighted voting involved in Boosting are considered to be biased. This is very similar to the human reasoning approach that people generally make decisions and judgments based on their previous experience without the guarantee that the decisions and judgments are absolutely right [1]. However, the frequency or the weight of a class can fairly be used to reflect the chance of the class being selected towards classifying a test instance, especially when the above conjecture concerning low quality training data cannot be proved in a reasonable way.

The impact of probabilistic voting on the Bagging and Boosting approaches was investigated experimentally in [4] by using 15 UCI data sets (see Table 4.2) and the results are shown in Table 4.3. In particular, the second and third columns (Random

Table 4.2 Data sets for ensemble learning experiments [4]

Name	Attribute types	Attributes	Instances	Classes
breast-cancer	Discrete	9	286	2
breast-w	Continuous	10	699	2
ecoli	Continuous	8	336	8
glass	Continuous	10	214	6
haberman	Mixed	4	306	2
heart-c	Mixed	76	920	4
heart-h	Mixed	76	920	4
heart-statlog	Continuous	13	270	2
hypothyroid	Mixed	30	3772	4
ionosphere	Continuous	34	351	2
iris	Continuous	5	150	3
labor	Mixed	17	57	2
sonar	Continuous	61	208	4
vote	Discrete	17	435	2
wine	Continuous	14	178	3

Forest I and II) indicate the results for Random Forests with majority voting [8] and Random Forests with probabilistic voting [4], respectively. Similarly, the fourth and fifth columns (Adaboost I and II) indicate the results for Adaboost with weighted voting [9] and Adaboost with probabilistic voting [4], respectively. Moreover, the Random Forests and Adaboost methods are the popular examples of Bagging and Boosting respectively in practical applications.

For both Bagging and Boosting, the experiment results show an improvement in classification accuracy with probabilistic voting. In particular, probabilistic voting can help both Random Forest and Adaboost effectively improve the overall accuracy of classification, except for the hypothyroid data set.

In addition, while the UCI data sets are a good benchmark for judging new approaches, they are known to be cleaner (i.e. contain fewer errors in the data) and more complete than data used in real-life applications, especially when considering the current vast volumes of data and the need to analyze data streams. Consequently, the benefits of probabilistic voting could be higher on this type of data where the assumptions of completeness and sample representativeness are rarely met; however, further experimentation is required to assess the benefits of probabilistic voting in this context.

Table 4.3 Accuracy of ensemble classification [4]

Dataset	Random forest I (%)	Random forest II (%)	Adaboost I (%)	Adaboost II (%)
breast-cancer	70	78	74	77
breast-w	95	96	94	96
ecoli	83	85	65	68
glass	69	80	45	52
haberman	68	74	72	78
heart-c	78	81	82	84
heart-h	82	87	79	80
heart-statlog	77	84	82	88
hypothyroid	98	98	95	95
ionosphere	89	94	89	90
iris	94	96	93	97
labor	90	95	91	95
sonar	76	83	75	83
vote	95	97	95	98
wine	94	98	88	91

References

1. H. Liu, A. Gegov, and M. Cocea. 2016. Nature and biology inspired approach of classification towards reduction of bias in machine learning. In *International Conference on Machine Learning and Cybernetics*, Jeju Island, South Korea, 10–13 July 2016, 588–593.
2. Man, K.F., K.S. Tang, and S. Kwong. 1996. Genetic algorithms: Concepts and applications. *IEEE Transactions on Industry Electronics* 43 (5): 519–534.
3. Lipowski, A., and D. Lipowska. 2012. Roulette-wheel selection via stochastic acceptance. *Physica A: Statistical Mechanics and its Applications* 391 (6): 2193–2196.
4. Liu, H., and M. Cocea. 2017. Granular computing based approach for classification towards reduction of bias in ensemble learning. *Granular Computing* 2 (3): 131–139.
5. Pedrycz, W., and S.-M. Chen. 2011. *Granular computing and intelligent systems: design with information granules of higher order and higher type*. Heidelberg: Springer.
6. Pedrycz, W., and S.-M. Chen. 2015. *Granular computing and decision-making: interactive and iterative approaches*. Heidelberg: Springer.
7. Pedrycz, W., and S.-M. Chen. 2015. *Information granularity, big data, and computational intelligence*. Heidelberg: Springer.
8. Breiman, L. 2001. Random forests. *Machine Learning* 45 (1): 5–32.
9. Y. Freund and R. E. Schapire. 1996. Experiments with a new boosting algorithm. In *Machine Learning: Proceedings of the Thirteenth International Conference*, Bari, Italy, 3–6 July 1996, 148–156.

Chapter 5
Fuzzy Classification Through Generative Multi-task Learning

Abstract In this chapter, we introduce the concepts of both generative learning and multi-task learning, and presents a proposed fuzzy approach for multi-task classification. We also discuss the advantages of fuzzy classification in the context of generative multi-task learning, in comparison with traditional classification in the context of discriminative single-task learning.

5.1 Overview of Generative Multi-task Learning

The purpose of generative multi-task learning is to remove the bias on mutual exclusion between different classes. As mentioned in Chap. 2, in traditional machine learning, different classes are mutually exclusive and each instance can only belong to one class. In other words, traditional learning approaches typically belong to discriminative learning, which aims at discriminating one class from other classes towards classifying an instance. Also, the task is learning to discriminate between classes towards identifying the class to be assigned to an unseen instance, so this type of learning is considered to be of single-task.

In contrast, in the context of generative learning, all classes are treated equally towards identifying the membership or non-membership of an instance to these classes. Also, in the context of multi-task learning, each class leads to an independent task of learning towards judging the membership or non-membership of an unseen instance to the class. As argued in [1], fuzzy approaches typically work in the way of generative learning and also involve multi-task learning. More details on fuzzy classification are given in Sect. 5.2.

5.2 Concepts of Fuzzy Classification

Fuzzy classification is a type of classification based on fuzzy logic. In this section, we describe concepts of fuzzy logic and its application in classification.

Fuzzy logic is an extension of deterministic logic, i.e. the truth value is ranged from 0 to 1 rather than a binary value. The theory of fuzzy logic is mainly aimed at turning a black and white problem into a grey problem [2]. In the context of set

H. Liu and M. Cocea, *Granular Computing Based Machine Learning*,
Studies in Big Data 35, https://doi.org/10.1007/978-3-319-70058-8_5

theory, deterministic logic is employed by crisp sets regarding the membership of an element to a set, which means that each element in a crisp set fully belongs to the set with no uncertainty. In contrast, fuzzy logic is employed by fuzzy sets, which indicates that each element in a fuzzy set may just partially belong to the set, i.e. the element belongs to the set to a certain degree referred to as fuzzy membership degree. In practice, the degree of a fuzzy membership can be measured by using a particular fuzzy membership functions such as trapezoid, triangular and Gaussian membership functions [3].

Fuzzy logic involves some logical operations that are slightly different from the operations used in deterministic logic such as conjunction, disjunction and negation. In terms of conjunction, the *min* function is used to get the smallest value among the values of the given fuzzy variables. For example, if a, b and c are three fuzzy variables with the fuzzy truth values of 0.4, 0.6 and 0.8 respectively, then $a \wedge b \wedge c = min(a, b, c) = 0.4$. For the same example, disjunction involves using the *max* function instead of the *min* function, i.e. $a \vee b \vee c = max(a, b, c) = 0.8$. In terms of negation, for the above example, $\neg a = 1 - a = 0.6$. More details on fuzzy operations can be found in [3].

There are three popular types of fuzzy rule based systems, namely Mamdani, Sugeno and Tsukamoto [3]. These three types of fuzzy rule based systems typically have the same form of antecedent (if part) but different forms of consequent (then part), and are typically used for regression problems, as the output from such systems is a real value. As we focus on classification, we present a modified form of rule consequent based on the above three types of fuzzy systems, in order to enable the output of discrete values and thus suit classification tasks.

The procedure of building a fuzzy rule based system involves fuzzification of numerical attributes and learning of fuzzy rules. In particular, for each input attribute, a fuzzy membership function is employed for transforming a numerical attribute into several linguistic attributes. The shape of membership functions we choose is trapezoid, since this shape is popularly used in practice [4]. In order to specify a trapezoidal membership function, four parameters (a, b, c, d) need to be estimated, as illustrated below and in Fig. 5.1:

$$f_T(x) = \begin{cases} 0, & \text{when } x \leq a \text{ or } x \geq d; \\ (x-a)/(b-a), & \text{when } a < x < b; \\ 1, & \text{when } b \leq x \leq c; \\ (d-x)/(d-c), & \text{when } c < x < d; \end{cases}$$

The learning of fuzzy rules could be based on the mixed fuzzy rule formation [6], which has been implemented in KNIME [7]. There are also other existing approaches for learning of fuzzy rules that can be found in [8, 9]. Following the completion of fuzzy rules learning, the five operations (fuzzification, application, implication, aggregation and defuzzification) are then executed in the classification stage. The following example [10] is given for illustration:

- Rule 1: If x_1 is 'Tall' and x_2 'Large' then y = 'Positive';
- Rule 2: If x_1 is 'Tall' and x_2 is 'Small' then y = 'Positive';

Fig. 5.1 Trapezoid fuzzy
membership function [5]

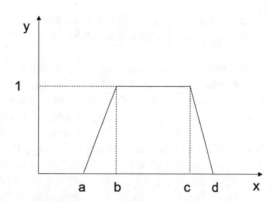

- Rule 3: If x_1 is 'Short' and x_2 is 'Large' then y = 'Negative';
- Rule 4: If x_1 is 'Short' and x_2 is 'Small' then y = 'Negative';

when $a = 1.3$, $b = 1.8$, $c = 1.8$ and $d = 1.8$ for the linguistic term 'Tall', $a = 2$, $b = 8$, $c = 8$ and $d = 8$ for the linguistic term 'Large', and if $x_1 = 1.425$ and $x_2 = 6.5$, then the following steps are executed.

Fuzzification:
Rule 1: $f_{Tall}(1.425) = 0.25$, $f_{Large}(6.5) = 0.75$;
Rule 2: $f_{Tall}(1.425) = 0.25$, $f_{Small}(6.5) = 0.25$;
Rule 3: $f_{Short}(1.425) = 0.75$, $f_{Large}(6.5) = 0.75$;
Rule 4: $f_{Short}(1.425) = 0.75$, $f_{Small}(6.5) = 0.25$;

In the fuzzification stage, the notation $f_{Tall}(1.425)$ represents the fuzzy membership degree of the numerical value '1.425' to the fuzzy linguistic term 'Tall'. Similarly, the notation $f_{Large}(6.5)$ represents the fuzzy membership degree of the numerical value '6.5' to the fuzzy linguistic term 'Large'. The fuzzification stage is aimed at mapping the numerical value of a variable to a membership degree to a particular fuzzy set.

Application:
Rule 1: $f_{Tall}(1.425) \wedge f_{Large}(6.5) = Min(0.25, 0.75) = 0.25$;
Rule 2: $f_{Tall}(1.425) \wedge f_{Small}(6.5) = Min(0.25, 0.25) = 0.25$;
Rule 3: $f_{Short}(1.425) \wedge f_{Large}(6.5) = Min(0.75, 0.75) = 0.75$;
Rule 4: $f_{Short}(1.425) \wedge f_{Small}(6.5) = Min(0.75, 0.25) = 0.25$;

In the application stage, the conjunction of the two fuzzy membership degrees respectively for the two variables 'x_1 and 'x_2' is aimed at deriving the firing strength of a fuzzy rule.

Implication:
Rule 1: $f_{Rule_1 \rightarrow Positive}(1.425, 6.5) = Min(0.25, 1) = 0.25$;
Rule 2: $f_{Rule_2 \rightarrow Positive}(1.425, 6.5) = Min(0.25, 1) = 0.25$;

Rule 3: $f_{Rule_3 \rightarrow Negative}(1.425, 6.5) = Min(0.75, 1) = 0.75;$
Rule 4: $f_{Rule_4 \rightarrow Negative}(1.425, 6.5) = Min(0.25, 1) = 0.25;$

In the implication stage, the firing strength of a fuzzy rule derived in the application stage can be used further to identify the membership degree of the input vector to the class 'Positive' or 'Negative', depending on the consequent of the fuzzy rule. For example, $f_{Rule_1 \rightarrow Positive}(1.425, 6.5) = 0.25$ indicates that the consequent of Rule 1 is the class 'Positive' and the input vector (1.425, 6.5) has the membership degree of 0.25 to the class 'Positive' through the inference from Rule 1. Similarly, $f_{Rule_3 \rightarrow Negative}(1.425, 6.5) = 0.75$ indicates that the consequent of Rule 3 is the class 'Negative' and the input vector (1.425, 6.5) has the membership degree of 0.75 to the class 'Negative' through the inference from Rule 3.

Aggregation:
$f_{Positive}(1.425, 6.5) = f_{Rule_1 \rightarrow Positive}(1.425, 6.5) \vee f_{Rule_2 \rightarrow Positive}(1.425, 6.5) = max(0.25, 0.25) = 0.25;$
$f_{Negative}(1.425, 6.5) = f_{Rule_3 \rightarrow Negative}(1.425, 6.5) \vee f_{Rule_4 \rightarrow Negative}(1.425, 6.5) = max(0.75, 0.25) = 0.75;$

In the aggregation stage, the membership degree of the input vector to the class label ('Positive' or 'Negative'), which is inferred from a rule, is compared with the other membership degrees inferred from the other rules, towards finding the maximum among all the membership degrees. For example, Rule 3 and Rule 4 both provide 'Negative' as the consequent and the membership degree values of the input vector (1.425, 6.5) to the class 'Negative' are 0.75 and 0.25, respectively, through the inference from the two rules. As the maximum of the fuzzy membership degrees is 0.75, the input vector (1.425, 6.5) is considered to have the membership degree of 0.75 to class 'Negative'. Similarly, the maximum of the membership degree values of the input vector to the class 'Positive' through the inference from Rule 1 and Rule 2 is 0.25, so the input vector (1.425, 6.5) is considered to have the membership degree of 0.25 to the class 'Positive'.

Defuzzification:
$f_{Negative}(1.425, 6.5) > f_{Positive}(1.425, 6.5) \rightarrow y = Negative$

In the defuzzification stage, the aim is to identify the class label to which the input vector has the highest membership degree. In this example, as the membership degree of the input vector (1.425, 6.5) to the class 'Negative' is 0.75, which is higher than the the membership degree (0.25) to the class 'Positive', the final output is 'Negative' towards classifying the unseen instance (1.425, 6.5, ?).

5.3 Granular Framework of Learning

As shown in Sect. 5.2, fuzzy rule based classification is typically done in the form of generative multi-task classification, by treating all classes equally and judging on each class independently regarding the degree of membership of an instance to the class.

In terms of learning fuzzy rules, most of existing approaches still follow the way of discriminative learning. Since traditional membership functions are typically defined subjectively by experts for fuzzifying numerical attributes, and experts generally assume that linguistic attributes transformed from the same numerical attribute are mutually exclusive, the sum of the values of these linguistic attributes (reflecting membership degree values) must be 1. For example, as shown in Fig. 5.2, the sum of the membership degree values to 'Low', 'Medium' and 'High' is always 1.

As argued in Chap. 2, the assumption of mutual exclusion does not always hold. In other words, the sum of the membership degree values could be higher than 1. For example, through an experimental study on the 'Anneal' data set retrieved from the UCI repository [11], several membership functions defined for the linguistic attributes transformed from the 'Carbon' attribute are shown in Fig. 5.3, which supports the above argumentation.

In order to achieve generative learning of fuzzy rules, we propose a granular framework of learning, as illustrated in Fig. 5.4.

In this framework, the training set is divided into n subsets, where n is the number of classes. Each of the training subset covers instances of a particular class, such that fuzzification of numerical attributes is specific for this class. In this context, for each class, a set of fuzzy rules are learned. In the fuzzy classification stage, each set of fuzzy rules is used toward judging the overall degree of membership of an unseen instance to the corresponding class. Finally, the membership degree values of the unseen instance to these classes can be used for the purpose of interpretation or defuzzification (towards a crisp classification).

The above description indicates that the proposed framework is designed to involve a learning task for each class and learning tasks for different classes are

Fig. 5.2 Traditional membership functions defined by experts

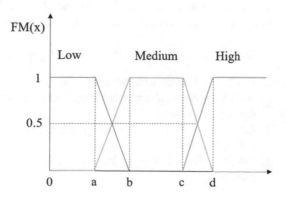

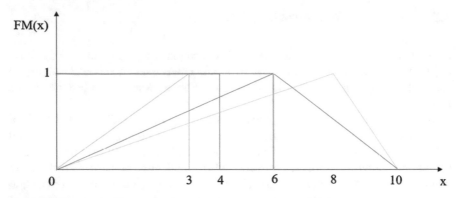

Fig. 5.3 Membership functions learned from anneal data on carbon attribute

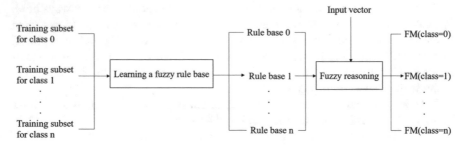

Fig. 5.4 Example of generative fuzzy rule learning

totally independent of each other, so the strategy of fuzzy rule learning follows both multi-task learning and generative learning.

In the setting of granular computing, the above framework involves explicitly both granulation and organization. In particular, granulation is to transform the class attribute into a number of binary attributes. Also, the training set is divided into a number of subsets, each of which corresponds to a particular class. Within each training subset, each numerical attribute is transformed into a number of linguistic attributes. Organization is to link the classes based on their relationships, such as mutual exclusion, correlation and independence, which will be discussed in detail in Sect. 5.4.

This framework involves three levels of granularity as follows:

1. Level 1: For each training subset, learning a membership function defined for each linguistic attribute transformed from a numerical attribute, and learning a set of fuzzy rules.
2. Level 2: For each class, learning to identify the fuzzy membership degree of an instance to the class.
3. Level 3: For all classes, learning to identify the relationships between these classes.

5.4 Discussion

Several existing approaches of fuzzy rule learning were used in [1] for generative multi-task classification showing empirically that an instance can belong to more than one class, due to the nature that fuzzy approaches consider each class to be assigned to an instance with a membership degree, i.e. the degree to which an instance belongs to each single class. In this context, the final classification is made by assigning an unseen instance the class with the highest fuzzy membership degree.

Since fuzzy approaches can show explicitly the fuzzy membership degree of an instance to each single class, we can observe if an instance has a fuzzy membership degree equal to or close to 1 for two or more classes. If the above phenomenon is frequently discovered, then it can strongly support the argumentation that an instance can belong to multiple classes. In addition, even if an instance appears to have a membership degree higher than 0.5 for at least two classes, it can still be considered that the instance weakly belongs to both of the two classes.

On the other hand, fuzzy approaches can be used to investigate if classes are mutually exclusive or not. In particular, if classes are mutually exclusive, then the sum of the fuzzy membership degrees for these classes should typically be equal to 1. This could have two different phenomena. One would show that the fuzzy membership degree of an instance is 1 to only one class and 0 to all the other classes. This phenomenon indicates that the instance fully belongs to one class only. The other one would show that an instance belongs to more than one class but the sum of the fuzzy membership degrees for these classes is equal to 1. This phenomenon indicates that the classes are mutually exclusive, but the instance is complex and belongs to different classes to different degrees. However, if the sum of the fuzzy membership degrees for these classes is greater than 1, then the classes are not mutually exclusive.

An experimental study was conducted in [1] by using 5 UCI data sets (see Table 5.1), for showing empirically how the fuzzy membership degree value of an instance to each single class can be checked and in what way it can be judged that an

Table 5.1 Data sets for fuzzy rule learning experiments [1]

Dataset	Attribute types	Attributes	Instances	Classes
anneal	Discrete, continuous	38	798	6
autos	Discrete, continuous	26	205	7
heart-c	Discrete, continuous	76	920	5
heart-h	Discrete, continuous	76	920	5
zoo	Discrete, continuous	18	101	7

Table 5.2 Results on anneal dataset [1]

ID	Class	1	2	3	4	5	U	Output
20	3	0	0	1	0	1	0	3
52	3	0	1	1	0	0	0	3
66	5	0	0.99	0	0	0.99	0	2
76	2	0	0.6	1	0	0	0	3
183	3	0	1	1	0	0	0	3
197	3	0	0	1	0	0	0	3
218	2	0	0.8	1	0	0	0	3
296	2	0	1	1	0	0	0	3
329	3	0	1	1	0	0	0	3
380	3	0	0	1	0	1	0	3
457	2	0	0.8	1	0	0	0	3
559	3	0	0.96	1	0	0	0	3
588	2	0	1	0.75	0	0	0	2
606	3	0	1	1	0	0	0	3
670	5	0	0	1	0	1	0	3
681	5	0	1	1	0	1	0	3
682	2	0	1	1	0	0	0	3
696	3	0	1	1	0	1	0	3
700	5	0	0	1	0	1	0	3
721	3	0	0.96	1	0	0	0	3
744	3	0	1	1	0	1	0	3
777	2	0	1	1	0	0	0	3
848	3	0.84	1	0.1	0	0	0	2
857	3	0	1	1	0	0	0	3
870	U	0	1	0.6	0	0	1	U

instance belongs to more than one class and if different classes are mutually exclusive or not. Experimental results on the Anneal data set are shown in Table 5.2.

Table 5.2 shows that 25 test instances (selected as representative examples from 200) are judged to belong to more than one class in accordance with the fuzzy membership degrees measured. In particular, three instances (681, 696 and 744) are judged to belong to three classes and the rest of the instances are judged to belong to two classes. Moreover, Table 5.2 shows 11 instances incorrectly classified according to traditional machine learning principles. However, looking at the columns 3 to 8, it can be noted that the above 11 instances may not be considered as incorrectly classified. For example, it can be seen for instance 66 that the original label assigned by experts is '5' and the predicted label is '2' but the instance has the membership degree of 0.99 to both classes (columns 4 and 7). Another example is instance 681—

it is assigned '3' as the predicted class label and experts assigned '5' as the original label, but according to the membership degree values the instance fully belongs to three classes ('2', '3' and '5'). Overall, most of the instances from this data set are assigned '3' as their predicted class label, but none of these cases can really be considered as incorrect classifications when looking at column 5 regarding the membership degree values for class '3'. In addition, this table shows to some extent the correlation between classes '2' and '3'. Results on several other data sets can be found in Appendix A.

The results shown in Table 5.2 indicate that an instance can belong to more than one class and that different classes may not be mutually exclusive and can even have correlations among each other.

As identified in Chap. 2, traditional discriminative single-task classification, which is dealt with by considering different classes to be mutually exclusive, may result in poor extendability of classifiers. In contrast, the generative multi-task classification is capable of addressing this issue by judging the membership of each instance to all classes. Thus, if a new class is added, a new learning task can be designed to deal with it without affecting the previous learning tasks on the other classes. Consequently, in this context, building a classifier is defined as a multiple learning task, in which each of the single learning tasks involves learning to judge the membership of an instance to a particular class, and these single learning tasks are generative on an independent or correlative basis. In this case, if a new class is added to the data set, then the classifier, which is learned from the original data set, can easily be extended by having another new single learning task on the updated data set. In other words, in the context of discriminative single-task classification, each classifier has only a single output, which is one or a subset of the predicted class labels. In contrast, in the context of generative multi-task classification, each classifier can have multiple outputs, each of which is corresponding to a particular class label. In practice, it is critical that a classifier can be easily extended in accordance with the dynamic update of a data set in terms of class labels.

On the other hand, as mentioned in Chap. 2, the assumption that different classes are mutually exclusive in a classification task does not always hold for real-life problems. For example, a book can belong to different subject areas. In fact, the nature of a classification problem is on prediction of the value of a discrete attribute. As introduced in [12], a discrete attribute can be specialized into different types, such as nominal, ordinal, string and categorical. For rating problems, the class attribute is of ordinal type. In this case, all the labels make up a whole enumeration so these labels need to be mutually exclusive. Also, classification tasks can be undertaken in practice for the purpose of decision making, which means to make a decision on the selection of one of the class labels. In this case, different classes also need to be mutually exclusive. When a classification task is undertaken for the purpose of categorization of items, it is very likely to occur that different classes have common instances, i.e. an instance can belong to multiple classes. In this case, it is not appropriate to consider that different classes are mutually exclusive. On the basis of the above statement, in the context of traditional discriminative single-task classification, some problems cannot be solved properly in practice and the nature of classifier learning is still learning to discriminate one class from other classes. However, in the context of generative multi-task classification, these problems can be addressed effectively by involving each class in a generative single learning task as part of multi-task learning. Also, the outcome of multi-task learning can be used as the basis for secondary learning towards identification of the relationships between different classes, such as generalization and aggregation.

In practice, the generative multi-task classification proposed in [1] can be achieved through both supervised and semi-supervised learning. For supervised learning, data labelling needs to be done by transforming the class attribute into several binary attributes, each of which is corresponding to a class label. In this way, experts need to judge on each class whether an instance belongs to it by assigning a membership degree value (0 or 1). For semi-supervised learning, data sets which have been previously used in traditional classification tasks, can be used again by transforming the class attribute into several binary attributes. In this way, the transformed data set would have all the binary attributes assigned truth values of 0 or 1. If an instance has its one of the binary attributes assigned 0, this would mean that the instance has not been labelled on the corresponding class. Otherwise, the instance would have been labelled on the corresponding class. On the basis of the transformed data set, fuzzy rule learning approaches can be used to measure the fuzzy membership degrees of each instance to each of the predefined class labels. Through cross-validation, each of the instances would be used in turn as a test instance to be measured on the extent to which the instance belongs to each single class. Finally, all these instances would have been assigned values of membership degree to each of the classes, which can be easily discretised to binary truth values.

References

1. H. Liu, M. Cocea, A. Mohasseb, and M. Bader. 2017. Transformation of discriminative single-task classification into generative multi-task classification in machine learning context. In *International Conference on Advanced Computational Intelligence*, Doha, Qatar, 4–6 February 2017, 66–73.
2. Zadeh, L. 2015. Fuzzy logic: A personal perspective. *Fuzzy Sets and Systems* 281: 4–20.
3. Ross, T. 2010. *Fuzzy Logic with Engineering Applications*. West Sussex: Wiley.
4. Chen, S.-M. 1996. A fuzzy reasoning approach for rule-based systems based on fuzzy logics. *IEEE Transactions on Systems, Man and Cybernetics - Part B: Cybernetics* 26 (5): 769–778.
5. H. Liu and M. Cocea. 2017. Fuzzy rule based systems for interpretable sentiment analysis. In *International Conference on Advanced Computational Intelligence*, Doha, Qatar, 4–6 February 2017, 129–136.
6. Berthold, M.R. 2003. Mixed fuzzy rule formation. *International Journal of Approximate Reasoning* 32: 67–84.
7. Berthold, M.R., B. Wiswedel, and T.R. Gabriel. 2013. Fuzzy logic in knime: Modules for approximate reasoning. *International Journal of Computational Intelligence Systems* 6 (1): 34–45.
8. Wang, L.-X., and J.M. Mendel. 1992. Generating fuzzy rules by learning from examples. *IEEE Transactions on Systems, Man and Cybernetics* 22 (6): 1414–1427.
9. Chen, S.-M., and L.-W. Lee. 2010. Fuzzy decision-making based on likelihood-based comparison relations. *IEEE Transactions on Fuzzy Systems* 18 (3): 613–628.

10. H. Liu and M. Cocea. Fuzzy information granulation towards interpretable sentiment analysis. *Granular Computing* 3 (1), In press.
11. M. Lichman. 2013. UCI machine learning repository. http://archive.ics.uci.edu/ml
12. Tan, P.-N., M. Steinbach, and V. Kumar. 2006. *Introduction to Data Mining*. New Jersey: Pearson Education.

Chapter 6
Multi-granularity Semi-random Data Partitioning

Abstract In this chapter, we introduce the concepts of semi-heuristic data partitioning, and present a proposed multi-granularity framework for semi-heuristic data partitioning. We also discuss the advantages of the proposed framework in terms of dealing with class imbalance and the sample representativeness issue, from granular computing perspectives.

6.1 Overview of Semi-random Data Partitioning

As introduced in Chap. 2, random data partitioning involves fully random sampling of training/test instances from an original data set. In the context of semi-random data partitioning, a population (data set) is divided into sub-populations (data subsets) and then simple random sampling is used within each sub-population for getting a sub-sample (strata).

In the context of machine learning, each class represents a sub-population and a training/test subset for a class represents a strata. Stratified sampling is typically used for improving the sample representiveness by reducing the data variability and thus reducing sampling error [1, 2]. An existing approach of semi-random partitioning is referred to as stratified sampling, which is popularly used in statistics [3].

As argued in Chap. 2, one of the purposes of adopting semi-random data partitioning is to avoid class imbalance through preserving the class distributions for training and test sets, especially for balanced or slightly imbalanced data sets. However, for this purpose, the stratified sampling technique needs to calculate the size of each strata based on its percentage of the total. For example, a data set has three classes with the distribution 40:40:20, the data partitioning needs to result in 70% of the data set for the training subset and 30% for the test subset.

While stratified sampling is adopted, Table 6.1 shows that each class needs to be given a probability for its instances to be selected into either the training set or the test set, i.e. it is needed to calculate the sampling probability for each class regarding the selection of its instances for the training/test set. This way aims to preserve the original class distribution in both the training and test sets but leads to higher computational complexity.

H. Liu and M. Cocea, *Granular Computing Based Machine Learning*,
Studies in Big Data 35, https://doi.org/10.1007/978-3-319-70058-8_6

Table 6.1 Sampling probability by stratified sampling [4]

Weight (%)	Probability for Class 1 (%)	Probability for Class 2 (%)	Probability for Class 3 (%)
Training set: 70	28	28	14
Test set: 30	12	12	6

On the basis of the above description, stratified sampling pays only attention to preserving the original class distribution by giving each class a sampling probability for its instances to be selected, without taking into account the balance between training and test samples.

6.2   Granular Framework of Learning

In this section, we present a multi-granularity framework proposed in [4] for effective control of the partitioning of a dataset into a training set and a test set. We also justify how the proposed approach can address the class imbalance and sample representativeness issues that can arise from random partitioning.

The multi-granularity framework for semi-random data partitioning is illustrated in Fig. 6.1. In particular, this framework involves three levels of granularity as outlined below:

1. Level 1: Data Partitioning is done randomly on the basis of the original data set towards getting a training set and a test set.
2. Level 2: The original data set is divided into a number of sub-sets, with each subset containing a class of instances. Within each subset (i.e. all instances with a particular class label), data partitioning into training and test sets is done randomly. The training and test sets for the whole dataset are obtained by merging all the training and test subsets, respectively.

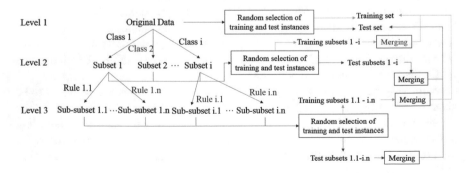

Fig. 6.1 Multi-granularity Framework for Semi-Random Data Partitioning [4]

3. Level 3: Based on the subsets obtained in Level 2, each of them is divided again
   into a number of sub-subsets, where each of the sub-subsets contains a subclass (of the corresponding class) of instances. The data partitioning is done
   randomly within each sub-subset. The training and test sets for the whole dataset
   are obtained by merging all the training and test sub-subsets, respectively.

In this multi-granularity framework, level 2 is aimed at addressing the class imbalance issue, i.e. to control the distribution of instances by class *within* the training and
test sets.

Level 3 is aimed at addressing the issue of sample representativeness, i.e. it is to
avoid the case that the training instances are highly dissimilar to the test instances
following the data partitioning.

In the setting of granular computing, the proposed framework involves explicitly
both granulation and organization. In particular, granulation is involved through the
operation that a data set is divided into a number of subsets and each subset is divided
into a training subset and a test subset (level 2), or further divided into sub-subsets and
then split into training and test sub-subsets (level 3). Also, organization is involved
by integrating the training subsets or sub-subsets into a whole training set, and the
test subsets or sub-subsets into a whole test set. In addition, in each level of the
granularity as shown in Fig. 6.1, a set of data is viewed as a granule, which also
has hierarchical relationships with sets of data (granules) located in other levels of
granularity.

Level 2 of the proposed multi-granularity framework is aimed at controlling effectively the selection of training/test instances towards avoiding the issue of class
imbalance, especially when the original data set is balanced. In particular, level 2 is
designed to ensure that for each class of instances a fixed percentage of the instances
would be included in the training/test set. For example, if we suppose that a data set
is divided into a training set and a test set in the ratio of 70:30, the strategy of semi-
random data partitioning involved in level 2 of the multi-granularity framework can
ensure that for each class of instances there would be 70% of the instances selected as
training instances and the rest of them selected as test instances. The above statement
can be proven as follows:

Let us suppose that a data set contains two classes (positive and negative) of
instances with the frequency distribution of $p : (1 - p)$, and the size of the data set
is m. Following data partitioning, the percentage of the training set is q whereas the
percentage of the test set is $1 - q$.

While the above strategy of semi-random data partitioning is taken, the following
steps would be involved:

1. Step 1: The data set is divided into two subsets respectively for the positive and
   negative classes, which results in mp positive instances and $m(1 - p)$ negative
   instances.
2. Step 2: Each class subset is partitioned into a training subset and a test subset. In
   particular, for the positive class, the size of the training subset is mpq and the size
   of the test subset is $mp(1 - q)$. Similarly, for the negative class, the size of the
   training subset is $m(1 - p)q$ and the size of the test subset is $m(1 - p)(1 - q)$.

Table 6.2 Sampling probability by semi-random partitioning

Weight (%)	Probability for Class 1 (%)	Probability for Class 2 (%)	Probability for Class 3 (%)
Training set: 70	70	70	70
Test set: 30	30	30	30

3. Step 3: The two training subsets resulting from Step 2 are merged into a whole training set and the frequency distribution between the positive and negative classes is $mpq : m(1 - p)q$, which is equivalent to $p : (1 - p)$, i.e. the original class distribution.

4. Step 4: The two test subsets resulting from Step 2 are merged into a whole test set and the frequency distribution between the positive and negative classes is $mp(1 - q) : m(1 - p)(1 - q)$, which is equivalent to $p : (1 - p)$, i.e. the original class distribution.

Thus, the procedure for level 2 ensures that the original class distribution for the whole data set is reflected within the training and test sets. The above proof, although demonstrated for a 2-class problem, also applies to multi-class classification problems, since the frequency distribution between different classes does not have any dependency on the number of classes as shown above.

In comparison with stratified sampling, the procedure for level 2 of the proposed multi-granularity framework only needs to divide a data set into subsets (each subset for a class) and then partition (in a fixed ratio) each subset into a training subset and a test subset, without the need to calculate the size of each training/test subset.

For the same example given in Sect. 6.1, Table 6.2 shows that it is not needed to calculate the sampling probability for each class regarding the selection of its instances for the training/test set. Instead, it is only needed to divide the original data set into n subsets, where n is the number of classes. For each subset corresponding to a class, it is just simply selecting an instance for the training/test set with 70/30% chance.

The above example indicates that the proposed semi-random partitioning pays more attention to balancing training and test sets by simply giving each instance 70/30% chance to be selected for the training/test set, in comparison with stratified sampling.

Level 3 of the proposed multi-granularity framework is aimed at controlling effectively the selection of training/test instances to ensure sample representativeness. In particular, the lack of sample representativess is likely to lead to overfitting, which means that a model performs well on the training data, but poorly on the test data. Thus, what the algorithm learns from the training data is not useful for the test data—something that is typically referred as a lack of generalization; in other words, the model is too specialized, i.e. it has learned from the training data very well, but cannot generalize this knowledge to other situations such as the ones in the test set.

To avoid this problem, the sample of data in the training set should be representative of the whole data, by ensuring that there is not a large dissimilarity between the training set and the test set. In order to avoid this dissimilarity, level 3 of the proposed multi-granularity framework is thus designed to involve grouping instances on the basis of their similarity to each other, and perform the partitioning within these groups, such that instances from the group will be present in both the training and the test sets.

6.3 Discussion

In this section, we show two parts of experimental results reported in [4]. In particular, the first part shows the performance of the proposed semi-random data partitioning in comparison with stratified sampling. The second part shows the performance of the proposed semi-random data partitioning in comparison with random data partitioning. Both experimental studies were conducted in [4] by using C4.5, Naive Bayes and K Nearest Neighbour. The characteristics of the data sets used are shown in Table 6.3

The results of the first experimental study are shown in Tables 6.4, 6.5 and 6.6. In these three tables, SS stands for stratified sampling and SR stands for semi-random partitioning.

Table 6.3 Data sets for semi-random data partitioning experiments [4]

Dataset	Feature types	#Attributes	#Instances	#Classes
anneal	Discrete, continuous	38	798	6
autos	Discrete, continuous	26	205	7
credit-a	Discrete, continuous	15	690	2
heart-stalog	Continuous	13	270	2
iris	Continuous	4	150	3
kr-vs-kp	Discrete	36	3196	2
labor	Discrete, continuous	17	57	2
segment	Continuous	19	2310	7
sonar	Continuous	60	208	2
tae	Discrete, continuous	6	151	3
vote	Discrete	16	435	2
wine	continuous	13	178	3

Table 6.4 Comparison with stratified sampling in terms of C4.5 performance [4]

Dataset			Class 1	Class 2	Class 3	Class 4	Class 5	Class 6	Class 7	Accuracy
aneal	SS	Precision	0.00	0.88	0.99	0.00	1.00	1.00		0.98
		Recall	0.00	1.00	0.98	0.00	1.00	1.00		
	SR	Precision	0.00	0.97	0.99	0.00	1.00	1.00		0.99
		Recall	0.00	1.00	1.00	0.00	1.00	1.00		
autos	SS	Precision	0.00	0.00	0.63	0.88	0.72	0.80	0.80	0.77
		Recall	0.00	0.00	0.71	0.7	0.81	0.80	1.00	
	SR	Precision	0.00	0.50	1.00	0.95	0.69	0.55	0.89	0.79
		Recall	0.00	1.00	0.57	0.95	0.69	0.60	1.00	
credit-a	SS	Precision	0.80	0.85						0.83
		Recall	0.82	0.83						
	SR	Precision	0.82	0.97						0.89
		Recall	0.97	0.83						
heart-statlog	SS	Precision	0.80	0.68						0.74
		Recall	0.71	0.78						
	SR	Precision	0.79	0.89						0.83
		Recall	0.93	0.69						
iris	SS	Precision	1.00	0.93	0.93					0.96
		Recall	1.00	0.93	0.93					
	SR	Precision	1.00	1.00	0.94					0.98
		Recall	1.00	0.93	1.00					
kr-vs-kp	SS	Precision	0.99	1.00						0.99
		Recall	1.00	0.99						
	SR	Precision	0.99	1.00						0.99
		Recall	1.00	0.99						
labor	SS	Precision	0.80	0.85						0.83
		Recall	0.67	0.92						
	SR	Precision	0.83	0.91						0.88
		Recall	0.83	0.91						
segment	SS	Precision	0.98	1.00	0.89	0.92	0.84	0.99	1.00	0.95
		Recall	0.98	1.00	0.90	0.92	0.83	1.00	0.99	
	SR	Precision	0.97	1.00	0.89	0.99	0.88	1.00	1.00	0.96
		Recall	0.97	1.00	0.89	0.94	0.93	1.00	1.00	
sonar	SS	Precision	0.65	0.72						0.68
		Recall	0.69	0.68						
	SR	Precision	0.81	0.87						0.84
		Recall	0.86	0.82						
tae	SS	Precision	0.40	0.39	0.46					0.41
		Recall	0.53	0.33	0.38					
	SR	Precision	0.55	0.67	0.55					0.57
		Recall	0.73	0.27	0.69					
vote	SS	Precision	0.94	0.98						0.96
		Recall	0.96	0.96						
	SR	Precision	0.97	0.94						0.96
		Recall	0.96	0.96						
wine	SS	Precision	1.00	0.96	0.93					0.96
		Recall	1.00	0.96	0.93					
	SR	Precision	1.00	0.91	1.00					0.96
		Recall	0.94	1.00	0.93					

Table 6.5 Comparison with stratified sampling in terms of NB performance [4]

Dataset			Class 1	Class 2	Class 3	Class 4	Class 5	Class 6	Class 7	Accuracy
aneal	SS	Precision	1.00	0.87	0.98	0.00	1.00	1.00		0.93
		Recall	1.00	0.87	0.92	0.00	1.00	0.50		
	SR	Precision	0.50	0.79	0.99	0.00	1.00	0.30		0.86
		Recall	1.00	1.00	0.82	0.00	1.00	0.92		
autos	SS	Precision	0.00	0.00	1.00	0.46	0.65	0.44	0.50	0.53
		Recall	0.00	0.00	0.14	0.6	0.81	0.40	0.38	
	SR	Precision	0.00	1.00	0.42	0.80	0.55	0.20	0.67	0.53
		Recall	0.00	1.00	0.71	0.40	0.69	0.20	0.75	
credit-a	SS	Precision	0.77	0.87						0.82
		Recall	0.85	0.79						
	SR	Precision	0.91	0.78						0.83
		Recall	0.67	0.95						
heart-statlog	SS	Precision	0.91	0.82						0.86
		Recall	0.84	0.89						
	SR	Precision	0.86	0.94						0.89
		Recall	0.96	0.81						
iris	SS	Precision	1.00	0.93	0.88					0.93
		Recall	1.00	0.87	0.93					
	SR	Precision	1.00	1.00	1.00					1.00
		Recall	1.00	1.00	1.00					
kr-vs-kp	SS	Precision	0.86	0.89						0.88
		Recall	0.91	0.84						
	SR	Precision	0.88	0.89						0.89
		Recall	0.91	0.87						
Labor	SS	Precision	1.00	0.86						0.89
		Recall	0.67	1.00						
	SR	Precision	1.00	1.00						1.00
		Recall	1.00	1.00						
segment	SS	Precision	1.00	1.00	0.68	0.53	0.49	1.00	1.00	0.75
		Recall	0.48	1.00	0.87	0.85	0.56	0.51	0.99	
	SR	Precision	0.79	1.00	0.57	0.90	0.43	0.95	1.00	0.80
		Recall	0.97	1.00	0.12	0.87	0.68	0.97	1.00	
sonar	SS	Precision	0.92	0.66						0.71
		Recall	0.41	0.97						
	SR	Precision	0.73	0.83						0.77
		Recall	0.83	0.73						
tae	SS	Precision	0.41	0.44	0.46					0.44
		Recall	0.47	0.53	0.31					
	SR	Precision	0.65	0.63	0.69					0.65
		Recall	0.73	0.67	0.56					
vote	SS	Precision	0.84	0.96						0.91
		Recall	0.94	0.89						
	SR	Precision	0.97	0.83						0.91
		Recall	0.88	0.96						
wine	SS	Precision	1.00	0.92	1.00					0.96
		Recall	0.94	1.00	0.93					
	SR	Precision	0.94	0.95	1.00					0.98
		Recall	0.97	1.00	1.00					

Table 6.6 Comparison with stratified sampling in terms of K-NN performance [4]

Dataset			Class 1	Class 2	Class 3	Class 4	Class 5	Class 6	Class 7	Accuracy
aneal	SS	Precision	0.00	0.63	0.86	0.00	0.75	0.83		0.83
		Recall	0.50	1.00	0.98	0.00	1.00	0.62		
	SR	Precision	1.00	0.90	0.99	0.00	1.00	0.69		0.96
		Recall	1.00	0.93	0.96	0.00	1.00	0.92		
autos	SS	Precision	0.00	0.00	0.00	0.33	0.54	0.11	0.00	0.32
		Recall	0.00	0.00	0.00	0.60	0.44	0.10	0.00	
	SR	Precision	0.00	0.00	0.71	0.58	0.55	0.50	0.67	0.58
		Recall	0.00	0.00	0.71	0.55	0.75	0.40	0.50	
credit-a	SS	Precision	0.66	0.71						0.69
		Recall	0.63	0.74						
	SR	Precision	0.91	0.88						0.89
		Recall	0.84	0.93						
heart-statlog	SS	Precision	0.64	0.54						0.59
		Recall	0.60	0.58						
	SR	Precision	0.84	0.88						0.85
		Recall	0.91	0.79						
iris	SS	Precision	1.00	1.00	0.94					0.98
		Recall	1.00	0.88	0.92					
	SR	Precision	1.00	0.88	1.00					0.96
		Recall	1.00	1.00	0.87					
kr-vs-kp	SS	Precision	0.52	0.00						0.52
		Recall	1.00	0.00						
	SR	Precision	0.94	0.97						0.96
		Recall	0.97	0.94						
labor	SS	Precision	0.86	1.00						0.94
		Recall	1.00	0.92						
	SR	Precision	1.00	0.92						0.94
		Recall	0.83	1.00						
segment	SS	Precision	0.96	1.00	0.88	0.93	0.89	1.00	1.00	0.95
		Recall	0.96	1.00	0.92	0.89	0.88	1.00	1.00	
	SR	Precision	0.96	1.00	0.85	0.99	0.87	0.96	1.00	0.95
		Recall	0.98	1.00	0.95	0.86	0.83	1.00	1.00	
sonar	SS	Precision	0.88	0.82						0.84
		Recall	0.76	0.91						
	SR	Precision	0.84	0.78						0.81
		Recall	0.72	0.88						
tae	SS	Precision	0.25	0.43	0.40					0.37
		Recall	0.20	0.40	0.50					
	SR	Precision	0.54	0.57	0.63					0.59
		Recall	0.47	0.53	0.75					
vote	SS	Precision	0.89	0.97						0.94
		Recall	0.96	0.93						
	SR	Precision	0.96	0.90						0.94
		Recall	0.94	0.94						
wine	SS	Precision	0.84	0.65	0.44					0.65
		Recall	0.94	0.50	0.53					
	SR	Precision	0.95	1.00	0.93					0.96
		Recall	1.00	0.90	1.00					

Table 6.4 shows that the proposed semi-random partitioning outperforms stratified sampling in 9 out of 12 cases, and the two approaches perform the same in the other 3 cases, in terms of overall accuracy of classification. Also, the proposed semi-random partitioning outperforms stratified sampling in terms of precision and recall with respect to each single class in most cases.

Table 6.5 shows that the proposed semi-random partitioning outperforms stratified sampling in 9 out of 12 cases, and the two approaches perform the same in 2 out of the other 3 cases, in terms of overall accuracy of classification. Also, the proposed semi-random partitioning outperforms stratified sampling in terms of precision and recall with respect to each single class in most cases.

Table 6.6 shows that the proposed semi-random partitioning outperforms stratified sampling in 7 out of 12 cases, and the two approaches perform the same in 3 out of the other 5 cases, in terms of overall accuracy of classification. Also, the proposed semi-random partitioning outperforms stratified sampling in terms of precision and recall with respect to each single class in most cases.

In the second experimental study, we show how random partitioning and semi-random partitioning can lead to different class frequency distributions for training and test sets, which can result in different impacts on the performance of learning by C4.5, Naive Bayes and K Nearest Neighbour. The results on class frequency distributions are shown in Tables 6.7, 6.8, 6.9 and 6.10.

Table 6.7 displays the original distribution of instances across classes for each dataset in terms of frequency (designated by #) and percentages (designated by %). For example, the anneal dataset (first row in Table 6.7) has 6 classes and in the original distribution class 1 has 8 instances (representing 1% of all instances), class 2 has 99 instances (representing 11% of the data) and so on. The same information is also displayed for the training and test sets used with the semi-random partitioning approach. The percentage numbers have been rounded to integers for ease of comparison. The loss of precision due to this rounding means that the sum across all classes may not be precisely 100%. Also, when the number of instances is low, a small difference in the number of instances may lead to a much bigger difference in the percentages values.

Tables 6.8, 6.9 and 6.10 show the original distribution, as well as the distribution within the training and test sets for C4.5, NB and K-NN, respectively. The original distribution was included in all tables for ease of comparison.

The results indicate that the random selection of data for training and test sets, leads to different effects on the distribution of instances across classes within the training and test sets, which are outlined below:

- For initially balanced datasets such as 'iris', 'segment' and 'tae', the random partitioning may lead to a loss of balance within the training and test sets; this loss can be observed for C4.5 on the 'iris' and 'tae' datasets, while for the 'segment' dataset, the variation is smaller; similarly, for NB, the loss of balance can be noticed for the 'iris' and 'tae' datasets, while for the 'segment' dataset, the variation is smaller, but more noticeable than for C4.5; for K-NN, a loss of balance can be

Table 6.7 Class frequency distribution with semi-random partitioning [4]

Data Set		Original distribution	Training set	Test set
anneal	#	8: 99: 684: 0: 67: 40	6: 69: 479: 0: 47: 28	2: 30: 205: 0: 20: 12
	%	1: 11: :76: 0: 7: 4	1: 11: 76: 0: 7: 4	1: 11: 76: 0: 7: 4
autos	#	0: 3: 22: 67: 54: 32: 27	0: 2: 15: 47: 38: 22: 19	0: 1: 7: 20: 16: 10: 8
	%	0: 1: 11: 33: 26: 16: 13	0: 1: 10: 33: 27: 15: 13	0: 2: 11: 32: 26: 16: 13
credit-a	#	307: 383	215: 268	92: 115
	%	44: 56	45: 55	44: 56
heart-statlog	#	150: 120	105: 84	45: 36
	%	56: 44	56: 44	56: 44
iris	#	50: 50: 50	35: 35: 35	15: 15: 15
	%	33: 33: 33	33: 33: 33	33: 33: 33
kr-vs-kp	#	1669: 1527	1168: 1069	501: 458
	%	52: 48	52: 48	52: 48
labor	#	20: 37	14: 26	6: 11
	%	35: 65	35: 65	35: 65
segment	#	330: 330: 330: 330: 330: 330: 330	231: 231: 231: 231: 231: 231: 231	99: 99: 99: 99: 99: 99: 99
	%	14: 14: 14: 14: 14: 14: 14	14: 14: 14: 14: 14: 14: 14	14: 14: 14: 14: 14: 14: 14
sonar	#	97: 111	68: 78	29: 33
	%	47: 53	47: 53	47: 53
tae	#	49: 50: 52	34: 35: 36	15: 15: 16
	%	32: 33: 34	32: 33: 34	33: 33: 35
vote	#	267: 168	187: 118	80: 50
	%	61: 39	61: 39	62: 38
wine	#	59: 71: 48	41: 50: 34	18: 21: 14
	%	33: 40: 27	33: 40: 27	34: 40: 26

observed for the 'tae' dataset, while for the iris dataset the imbalance is very small and for the 'segment' dataset the variation is small and similar to the variation for C4.5.

- For slightly imbalanced datasets, the random partitioning may lead to a more balanced distribution in the training set, but a more imbalanced one in the test set, i.e. for C4.5., 'heart-statlog'; for NB, labor and vote; for K-NN, 'credit-a', 'labor' and 'sonar'. Sometimes, the imbalance in the test set may mean that the majority class from the training set becomes minority class in the test set—this occurs only for one dataset, i.e. 'sonar' with K-NN, which is probably due to the fact that the distribution in this dataset is very close to perfect balance (47:53).
- For slightly imbalanced datasets, the random partitioning may lead to a more balanced distribution in the test set, but a more imbalanced distribution in the training set, i.e. for C4.5, 'kr-vs-kp' and 'labor' by C4.5; for NB, 'heart-statlog'.

Table 6.8 C4.5: class frequency distribution in training and test sets for random partitioning [4]

Dataset		Original distribution	Training set	Test set
anneal	#	8: 99: 684: 0: 67: 40	7: 73: 483: 0: 39: 27	1: 26: 201: 0: 28: 13
	%	1: 11: 76: 0: 7: 4	1: 12: 77: 0: 6: 4	0: 10: 75: 0: 10: 5
autos	#	0: 3: 22: 67: 54: 32: 27	0: 3: 17: 41: 43: 23: 17	0: 0: 5: 26: 11: 9: 10
	%	0: 1: 11: 33: 26: 16: 13	0: 2: 12: 28: 30: 16: 12	0: 0: 8: 43: 18: 15: 16
credit-a	#	307: 383	211: 272	96: 111
	%	44: 56	44: 56	46: 54
heart-statlog	#	150: 120	99: 90	51: 30
	%	56: 44	52: 48	63: 37
iris	#	50: 50: 50	38: 30: 37	12: 20: 13
	%	33: 33: 33	36: 29: 35	27: 44: 29
kr-vs-kp	#	1669: 1527	1196: 1041	473: 486
	%	52: 48	53: 47	49: 51
labor	#	20: 37	13: 27	7: 10
	%	35: 65	33: 68	41: 59
segment	#	330: 330: 330: 330: 330: 330: 330	223: 223: 230: 239: 242: 229: 231	107: 107: 100: 91: 88: 101: 99
	%	14: 14: 14: 14: 14: 14: 14	14: 14: 14: 15: 15: 14: 14	15: 15: 14: 13: 13: 15: 14
sonar	#	97: 111	62: 84	35: 27
	%	47: 53	42: 58	56: 44
tae	#	49: 50: 52	34: 32: 40	15: 18: 12
	%	32: 33: 34	32: 30: 38	33: 40: 27
vote	#	267: 168	186: 119	81: 49
	%	61: 39	61: 39	62: 38
wine	#	59: 71: 48	42: 55: 28	17: 16: 20
	%	33: 40: 27	34: 44: 22	32: 30: 38

For two of these, C4.5—'kr-vs-kp' and NB—'heart-statlog', in the test set, the majority class is reversed in comparison with the training set.

- For slightly imbalanced datasets, the random partitioning may lead to both the training and test sets to become more imbalanced, with a different class being the majority class in the training and test sets; for example, in the 'sonar' dataset with C4.5, class 2 is the majority class in the training set, while class 1 is the majority class in the test set. This situation occurs on the 'sonar' dataset for C4.5 and NB, and on the 'wine' dataset for all algorithms (C4.5, NB and K-NN).
- For the datasets with a high number of classes and an imbalanced distribution, e.g. anneal and autos, the random partitioning may preserve the original distribution for some classes, while for others, there is an imbalance in the training set, the test set or both, i.e. the 'autos' dataset for all algorithms (C4.5, NB, and K-NN); sometimes, the majority class in the training set is no longer the majority class in

Table 6.9 NB: class frequency distribution in training and test sets for random partitioning [4]

Dataset		Original distribution	Training set	Test set
anneal	#	8: 99: 684: 0: 67: 40	4: 67: 484: 0: 44: 30	4: 32: 200: 0: 23: 10
	%	1: 11: 76: 0: 7: 4	1: 11: 77: 0: 7: 5	1: 12: 74: 0: 9: 4
autos	#	0: 3: 22: 67: 54: 32: 27	0: 2: 15: 45: 39: 23: 20	0: 1: 7: 22: 15: 9: 7
	%	0: 1: 11: 33: 26: 16: 13	0: 1: 10: 31: 27: 16: 14	0: 2: 11: 36: 25: 15: 11
credit-a	#	307: 383	216: 267	91: 116
	%	44: 56	45: 55	44: 56
heart-statlog	#	150: 120	111: 78	39: 42
	%	56: 44	59: 41	48: 52
iris	#	50: 50: 50	37: 31: 37	13: 19: 13
	%	33: 33: 33	35: 30: 35	29: 42: 29
kr-vs-kp	#	1669: 1527	1164: 1073	505: 454
	%	52: 48	52: 48	53: 47
labor	#	20: 37	16: 24	4: 13
	%	35: 65	40: 60	24: 76
segment	#	330: 330: 330: 330: 330: 330: 330	245: 228: 229: 220: 245: 218: 232	85: 102: 101: 110: 85: 112: 98
	%	14: 14: 14: 14: 14: 14: 14	15: 14: 14: 14: 15: 13: 14	12: 15: 15: 16: 12: 16: 14
sonar	#	97: 111	60: 86	37: 25
	%	47: 53	41: 59	60: 40
tae	#	49: 50: 52	34: 31: 41	15: 19: 11
	%	32: 33: 34	32: 29: 39	33: 42: 24
vote	#	267: 168	183: 122	84: 46
	%	61: 39	60: 40	65: 35
wine	#	59: 71: 48	48: 43: 34	11: 28: 14
	%	33: 40: 27	38: 34: 27	21: 53: 26

the test set, e.g. for C4.5—'autos', class 5 is the majority class in the training set, while class 4 is the majority class in the test set (as well as the original dataset). For the anneal dataset, the distribution changes slightly, but the majority of the changes are less than 2%—for this reason we consider that the distribution for this dataset with all algorithms is very similar to the original distribution.

- For all datasets, the random partitioning may lead to a very similar distribution in the training and test sets as in the original dataset. i.e. for C4.5, 'anneal', 'credit-a' and 'vote'; for NB, 'anneal', 'credit-a' and 'kr-vs-kp'; for K-NN, 'anneal', 'heart-statlog', 'kr-vs-kp' and 'vote'.

In terms of the impacts on the performance of learning by C4.5, Naive Bayes and K Nearest Neighbour, the results reported in [4] are shown in Tables 6.11, 6.12 and 6.13 and in Appendix B, and could be summarized as follows: we noticed that the distribution of classes within the training and test sets, has an effect on the

Table 6.10 K-NN: class frequency distribution in training and test sets for random partitioning [4]

Dataset		Original distribution	Training set	Test set
anneal	#	8: 99: 684: 0: 67: 40	4: 64: 484: 0: 50: 27	4: 35: 200: 0: 17: 13
	%	1: 11: 76: 0: 7: 4	1: 10: 77: 0: 8: 4	1: 13: 74: 0: 6: 5
autos	#	0: 3: 22: 67: 54: 32: 27	0: 3: 16: 49: 38: 21: 17	0: 0: 6: 18: 16: 11: 10
	%	0: 1: 11: 33: 26: 16: 13	0: 2: 11: 34: 26: 15:12	0: 0: 10: 30: 26: 18: 16
credit-a	#	307: 383	224: 259	83: 124
	%	44: 56	46: 54	40: 60
heart-statlog	#	150: 120	106: 83	44: 37
	%	56: 44	56: 44	54: 46
iris	#	50: 50: 50	35: 34: 36	15: 16: 14
	%	33: 33: 33	33: 32: 34	33: 36: 31
kr-vs-kp	#	1669: 1527	1177: 1060	492: 467
	%	52: 48	53: 47	51: 49
labor	#	20: 37	15: 25	5: 12
	%	35: 65	38: 63	29: 71
segment	#	330: 330: 330: 330: 330: 330: 330	220: 223: 231: 238: 241: 234: 230	110: 107: 99: 92: 89: 96: 100
	%	14: 14: 14: 14: 14: 14: 14	14: 14: 14: 15: 15: 14: 14	16: 15: 14: 13: 13: 14: 14
sonar	#	97: 111	64: 82	33: 29
	%	47: 53	44: 56	53: 47
tae	#	49: 50: 52	33: 35: 38	16: 15: 14
	%	32: 33: 34	31: 33: 36	36: 33: 31
vote	#	267: 168	186: 119	81: 49
	%	61: 39	61: 39	62: 38
wine	#	59: 71: 48	34: 55: 36	25: 16: 12
	%	33: 40: 27	27: 44: 29	47: 30: 23

performance results. In particular, there is an association between a larger number of instances in the training set and a higher recall. and between a larger number of instances in the test set and a higher precision. A higher number of instances in the training set can mean more opportunities for learning, and, thus, better knowledge of a particular class, which explains the higher recall. For a good performance, however, recall needs to be balanced with precision, i.e. ensure that the model can distinguish a particular class from the other classes; in other words, a low precision means that instances of a particular class are wrongly labeled with another class(es). This is more likely to be influenced by the distribution among classes, than the distribution of a class between the training and the test set, as the balance between classes in the training set has an influence on the capacity to learn to distinguish between classes (which is why class imbalance is known to lead to poor performance). This is supported by the fact that the semi-random partitioning results are more balanced

Table 6.11 C4.5 performance on accuracy, precision and recall [4]

Dataset			Class 1	Class 2	Class 3	Class 4	Class 5	Class 6	Class 7	Accuracy
aneal	R	Precision	0.00	0.96	0.99	0.00	1.00	1.00		0.99
		Recall	0.00	1.00	1.00	0.00	1.00	0.85		
	SR	Precision	0.00	0.97	0.99	0.00	1.00	1.00		0.99
		Recall	0.00	1.00	1.00	0.00	1.00	1.00		
autos	R	Precision	0.00	0.00	1.00	0.92	0.50	0.78	0.75	0.77
		Recall	0.00	0.00	0.80	0.85	0.73	0.78	0.60	
	SR	Precision	0.00	0.50	1.00	0.95	0.69	0.55	0.89	0.79
		Recall	0.00	1.00	0.57	0.95	0.69	0.60	1.00	
credit-a	R	Precision	0.82	0.90						0.86
		Recall	0.90	0.83						
	SR	Precision	0.82	0.97						0.89
		Recall	0.97	0.83						
heart-statlog	R	Precision	0.97	0.67						0.81
		Recall	0.73	0.97						
	SR	Precision	0.79	0.89						0.83
		Recall	0.93	0.69						
iris	R	Precision	1.00	0.94	0.81					0.91
		Recall	0.92	0.85	1.00					
	SR	Precision	1.00	1.00	0.94					0.98
		Recall	1.00	0.93	1.00					
kr-vs-kp	R	Precision	0.99	0.99						0.99
		Recall	0.99	0.99						
	SR	Precision	0.99	1.00						0.99
		Recall	1.00	0.99						
labor	R	Precision	0.67	0.64						0.65
		Recall	0.29	0.90						
	SR	Precision	0.83	0.91						0.88
		Recall	0.83	0.91						
segment	R	Precision	0.97	0.98	0.92	0.99	0.84	1.00	0.99	0.96
		Recall	0.99	1.00	0.91	0.88	0.90	1.00	1.00	
	SR	Precision	0.97	1.00	0.89	0.99	0.88	1.00	1.00	0.96
		Recall	0.97	1.00	0.89	0.94	0.93	1.00	1.00	
sonar	R	Precision	0.85	0.79						0.82
		Recall	0.83	0.81						
	SR	Precision	0.81	0.87						0.84
		Recall	0.86	0.82						
tae	R	Precision	0.50	0.56	0.60					0.56
		Recall	0.40	0.56	0.75					
	SR	Precision	0.55	0.67	0.55					0.57
		Recall	0.73	0.27	0.69					
vote	R	Precision	0.95	0.96						0.95
		Recall	0.98	0.92						
	SR	Precision	0.97	0.94						0.96
		Recall	0.96	0.96						
wine	R	Precision	0.94	0.83	0.94					0.91
		Recall	0.94	0.94	0.85					
	SR	Precision	1.00	0.91	1.00					0.96
		Recall	0.94	1.00	0.93					

Table 6.12 NB performance on accuracy, precision and recall [4]

Dataset			Class 1	Class 2	Class 3	Class 4	Class 5	Class 6	Class 7	Accuracy
aneal	R	Precision	0.66	0.76	0.99	0.00	1.00	0.24		0.84
		Recall	0.50	1.00	0.79	0.00	1.00	1.00		
	SR	Precision	0.50	0.79	0.99	0.00	1.00	0.30		0.86
		Recall	1.00	1.00	0.82	0.00	1.00	0.92		
autos	R	Precision	0.00	1.00	0.30	0.50	0.59	0.50	0.67	0.52
		Recall	0.00	1.00	0.43	0.32	0.87	0.44	0.57	
	SR	Precision	0.00	1.00	0.42	0.80	0.55	0.20	0.67	0.53
		Recall	0.00	1.00	0.71	0.40	0.69	0.20	0.75	
credit-a	R	Precision	0.89	0.79						0.82
		Recall	0.68	0.93						
	SR	Precision	0.91	0.78						0.83
		Recall	0.67	0.95						
heart-statlog	R	Precision	0.77	0.94						0.84
		Recall	0.95	0.74						
	SR	Precision	0.86	0.94						0.89
		Recall	0.96	0.81						
iris	R	Precision	1.00	1.00	0.87					0.96
		Recall	1.00	0.89	1.00					
	SR	Precision	1.00	1.00	1.00					1.00
		Recall	1.00	1.00	1.00					
kr-vs-kp	R	Precision	0.87	0.89						0.88
		Recall	0.91	0.85						
	SR	Precision	0.88	0.89						0.89
		Recall	0.91	0.87						
labor	R	Precision	0.75	0.92						0.88
		Recall	0.75	0.92						
	SR	Precision	1.00	1.00						1.00
		Recall	1.00	1.00						
segment	R	Precision	0.76	1.00	0.69	0.91	0.41	0.98	1.00	0.81
		Recall	0.99	1.00	0.18	0.83	0.71	0.97	0.99	
	SR	Precision	0.79	1.00	0.57	0.90	0.43	0.95	1.00	0.80
		Recall	0.97	1.00	0.12	0.87	0.68	0.97	1.00	
sonar	R	Precision	0.77	0.79						0.77
		Recall	0.89	0.60						
	SR	Precision	0.73	0.83						0.77
		Recall	0.83	0.73						
tae	R	Precision	0.63	0.50	0.25					0.47
		Recall	0.80	0.26	0.36					
	SR	Precision	0.65	0.63	0.69					0.65
		Recall	0.73	0.67	0.56					
vote	R	Precision	0.96	0.81						0.90
		Recall	0.88	0.93						
	SR	Precision	0.97	0.83						0.91
		Recall	0.88	0.96						
wine	R	Precision	1.00	1.00	1.00					1.00
		Recall	1.00	1.00	1.00					
	SR	Precision	0.94	0.95	1.00					0.98
		Recall	0.97	1.00	1.00					

Table 6.13 K-NN performance on accuracy, precision and recall [4]

Dataset			Class 1	Class 2	Class 3	Class 4	Class 5	Class 6	Class 7	Accuracy
aneal	R	Precision	0.67	0.95	0.97	0.00	1.00	0.89		0.96
		Recall	0.50	1.00	0.98	0.00	1.00	0.62		
	SR	Precision	1.00	0.90	0.99	0.00	1.00	0.69		0.96
		Recall	1.00	0.93	0.96	0.00	1.00	0.92		
autos	R	Precision	0.00	0.00	0.67	0.60	0.65	0.29	1.00	0.52
		Recall	0.00	0.00	0.33	0.67	0.69	0.36	0.30	
	SR	Precision	0.00	0.00	0.71	0.58	0.55	0.50	0.67	0.58
		Recall	0.00	0.00	0.71	0.55	0.75	0.40	0.50	
credit-a	R	Precision	0.85	0.88						0.87
		Recall	0.82	0.90						
	SR	Precision	0.91	0.88						0.89
		Recall	0.84	0.93						
heart-statlog	R	Precision	0.71	0.70						0.70
		Recall	0.77	0.62						
	SR	Precision	0.84	0.88						0.85
		Recall	0.91	0.79						
iris	R	Precision	1.00	0.93	0.87					0.93
		Recall	1.00	0.88	0.92					
	SR	Precision	1.00	0.88	1.00					0.96
		Recall	1.00	1.00	0.87					
kr-vs-kp	R	Precision	0.93	0.98						0.95
		Recall	0.98	0.92						
	SR	Precision	0.94	0.97						0.96
		Recall	0.97	0.94						
labor	R	Precision	1.00	0.92						0.94
		Recall	0.80	1.00						
	SR	Precision	1.00	0.92						0.94
		Recall	0.83	1.00						
segment	R	Precision	0.98	1.00	0.87	0.94	0.80	0.98	1.00	0.94
		Recall	0.97	1.00	0.91	0.88	0.82	1.00	0.99	
	SR	Precision	0.96	1.00	0.85	0.99	0.87	0.96	1.00	0.95
		Recall	0.98	1.00	0.95	0.86	0.83	1.00	1.00	
sonar	R	Precision	0.86	0.74						0.79
		Recall	0.73	0.86						
	SR	Precision	0.84	0.78						0.81
		Recall	0.72	0.88						
tae	R	Precision	0.44	0.36	0.50					0.44
		Recall	0.50	0.27	0.57					
	SR	Precision	0.54	0.57	0.63					0.59
		Recall	0.47	0.53	0.75					
vote	R	Precision	0.97	0.87						0.93
		Recall	0.91	0.96						
	SR	Precision	0.96	0.90						0.94
		Recall	0.94	0.94						
wine	R	Precision	1.00	1.00	0.92					0.98
		Recall	1.00	0.93	1.00					
	SR	Precision	0.95	1.00	0.93					0.96
		Recall	1.00	0.90	1.00					

in terms of precision and recall, while the random partitioning with imbalanced class distribution in the training set, as well as imbalance across the training and test sets, tend to have one of two combinations: (a) high precision and low recall, or (b) low recall and high precision.

The results also indicate that the class distribution within the training set has more influence on the performance than the class distribution within the test sct. On the other hand, the distribution within the test set still requires consideration to accurately assess the performance of a model. For example, a small test sample may not sufficiently test the knowledge learned for a particular class—in an extreme situation it may mean that knowledge is not tested at all. These aspects can be easily controlled with our proposed partitioning method.

Overall, the experimental results indicate that the adoption of the strategy of semi-random data partitioning involved in level 2 of the multi-granularity framework presented in Sect. 6.2 achieves effective control of the selection of training/test instances, towards avoiding the case of class imbalance in both training and test sets, especially when data sets are originally balanced or slightly imbalanced.

The results also showed situations when the random and semi-random partitioning led to the same distribution, but different results. We believe these are likely to be explained by the sample representativeness issue, which will be addressed in future work with experiments on Level 3 of the propose multi-granularity framework.

References

1. Esfahani, M.S., and E.R. Dougherty. 2014. Effect of separate sampling on classification accuracy. *Bioinformatics* 30 (2): 242–250.
2. K. Lang, E. Liberty, and K. Shmakov. 2016. Stratified sampling meets machine learning. In *Proceedings of the 33rd International Conference on Machine Learning*. New York: JMLR.org, 2320–2329.
3. C.-E. Särndal B. Swensson, and J. Wretman. 1992. Model Assisted Survey Sampling. New York: Springer.
4. H. Liu and M. Cocea. Semi-random partitioning of data into training and test sets in granular computing context. *Granular Computing*, 2 (4) 2017.

Chapter 7
Multi-granularity Rule Learning

Abstract In this chapter, we introduce concepts of rule learning and review existing methods for identifying their limitations. Based on the review, we present a proposed multi-granularity framework of rule learning, towards advancing the learning performance and improving the quality of each single rule learned. Furthermore, we discuss the advantages of multi-granularity rule learning, in comparison with traditional rule learning.

7.1 Overview of Rule Learning

As introduced in [1–3], rule learning can be achieved following two approaches, namely, divide and conquer [4], and separate and conquer [5]. The former is aimed at learning a set of rules represented in the form of a decision tree, such as ID3 [6], C4.5 [4] and Classification and Regression Tree (CART) [7], whereas the latter is aimed at learning a set of if-then rules directly from training instances, such as Prism [8] and Information Entropy Based Rule Generation(IEBRG) [9].

The divide and conquer approach is also known as Top-Down Induction of Decision Trees (TDIDT), due to the fact that rules learned through using this approach are represented in the form of decision trees and that the induction (learning) procedure is from general to specific like the top-down approach in computer science. The procedure of the TDIDT approach is illustrated in Fig. 7.1.

The separate and conquer approach is also known as the covering approach, due to the fact that this approach typically involves learning a set of rules sequentially. The aim of this approach is to learn a rule that covers some instances of the same class, and then to start iteratively learning the next rule from the rest of the training instances. These instances should not have been covered by the rules that are learned previously. In other words, all the above instances covered by the previously learned rules need to have been deleted from the current training subset. The procedure of the covering approach is illustrated in Fig. 7.2.

As argued in [5, 8], the TDIDT approach may lead to the production of a complex tree (equivelant to a large number of complex rules) and even the replicated sub-tree problem (see Fig. 7.3). These issues may not only result in overfitting of training

© Springer International Publishing AG 2018

H. Liu and M. Cocea, *Granular Computing Based Machine Learning*,

Studies in Big Data 35, https://doi.org/10.1007/978-3-319-70058-8_7

Input: A set of training instances, attribute A_i, where i is the index of the attribute A, value V_j, where j is the index of the value V
Output: A decision tree.
    if the stopping criterion is satisfied **then**
        create a leaf that corresponds to all remaining training instances
    else
        choose the best (according to some heuristics) attribute A_i
        label the current node with A_i
        for each value V_j of the attribute A_i **do**
        label an outgoing edge with value V_j
        recursively build a subtree by using a corresponding subset of training   instances
        end for
    end if

Fig. 7.1   Procedure of TDIDT approach [10, 11]

Input: A set of training instances
Output: An ordered set of rules
    while training set is not empty **do**
        generate a single rule from the training set
        delete all instances covered by this rule
        if the generated rule is not good **then**
            generate the majority rule **and** empty the training set
        end if
    end while

Fig. 7.2   Procedure of covering approach [10, 11]

data but also affect the interpretability of a decision tree [12]. In order to address the above issues, researchers have been motivated to develop methods of separate and conquer rule learning, towards production of a smaller number of simpler rules. Some related work can be found in [5, 8].

In this chapter, we focus on separate and conquer rule learning (the covering approach), so we illustrate this approach through the IEBRG algorithm, by using the Weather data set as shown in Table 7.1.

According to Table 7.1, a frequency table can be derived for each attribute, i.e. we have four frequency tables for the four attributes: Outlook (Table 7.2), Temperature (Table 7.3), Humidity (Table 7.4) and Windy (Table 7.5).

Based on the frequency tables, the value of conditional entropy for each attribute-value pair of each attribute can be calculated. We display these here for ease of explanation – in the normal course of the algorithm the entropy values would be calculated when needed, not in advance.

According to Table 7.2, we can derive the conditional entropy value for each of the three values of attribute 'Outlook'.

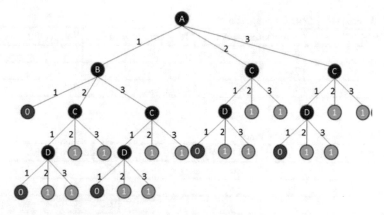

Fig. 7.3 Cendrowska's replicated sub-tree problem [1, 12]

Table 7.1 Weather dataset

Outlook	Temperature	Humidity	Windy	Play?
Sunny	Hot	High	False	No
Sunny	Hot	High	True	No
Overcast	Hot	High	False	Yes
Rain	Mild	High	False	Yes
Rain	Cool	Normal	False	Yes
Rain	Cool	Normal	True	No
Overcast	Cool	Normal	True	Yes
Sunny	Mild	High	False	No
Sunny	Cool	Normal	False	Yes
Rain	Mild	Normal	False	Yes
Sunny	Mild	Normal	True	Yes
Overcast	Mild	High	True	Yes
Overcast	Hot	Normal	False	Yes
Rain	Mild	High	True	No

Table 7.2 Frequency table for outlook

Class label	Outlook = sunny	Outlook = overcast	Outlook = rain
Yes	2	4	3
No	3	0	2
Total	5	4	5

Table 7.3 Frequency table for temperature

Class label	Temperature = hot	Temperature = mild	Temperature = cool
Yes	2	4	3
No	2	2	1
Total	4	6	4

Table 7.4 Frequency table for humidity

Class label	Humidity = high	Humidity = normal
Yes	3	6
No	4	1
Total	7	7

Table 7.5 Frequency table for windy

Class label	Windy = true	Windy = false
Yes	3	6
No	3	2
Total	6	8

$E(Outlook = sunny) = -P(Class = Yes|Outlook = sunny) \times \log_2$
$P(Class = Yes|Outlook = sunny) - P(Class = No|Outlook = sunny) \times$
$\log_2 P(Class = No|Outlook = sunny) = -\frac{2}{5} \times \log_2 \frac{2}{5} - \frac{3}{5} \times \log_2 \frac{3}{5}$

$E(Outlook = overcast) = -P(Class = Yes|Outlook = overcast) \times \log_2$
$P(Class = Yes|Outlook = overcast) - P(Class = No|Outlook = overcast) \times \log_2 P(Class = No|Outlook = overcast) = -\frac{4}{4} \times \log_2 \frac{4}{4} - \frac{0}{4} \times \log_2 \frac{0}{4} = 0$

$E(Outlook = rain) = -P(Class = Yes|Outlook = rain) \times \log_2 P(Class = Yes|Outlook = rain) - P(Class = No|Outlook = rain) \times \log_2 P(Class = No|Outlook = rain) = -\frac{3}{5} \times \log_2 \frac{3}{5} - \frac{2}{5} \times \log_2 \frac{2}{5}$

According to Table 7.3, we can derive the conditional entropy value for each of the three values of attribute 'Temperature'.

$E(Temperature = hot) = -P(Class = Yes|Temperature = hot) \times \log_2$
$P(Class = Yes|Temperature = hot) - P(Class = No|Temperature = hot)$
$\times \log_2 P(Class = No|Temperature = hot) = -\frac{1}{2} \times \log_2 \frac{1}{2} - \frac{1}{2} \times \log_2 \frac{1}{2} = 1$

$E(Temperature = mild) = -P(Class = Yes|Temperature = mild) \times \log_2$
$P(Class = Yes|Temperature = mild) - P(Class = No|Temperature = mild) \times \log_2 P(Class = No|Temperature = mild) = -\frac{2}{3} \times \log_2 \frac{2}{3} - \frac{1}{3} \times \log_2 \frac{1}{3}$

$E(Temperature = cool) = -P(Class = Yes|Temperature = cool) \times \log_2$
$P(Class = Yes|Temperature = cool) - P(Class = No|Temperature = cool) \times \log_2 P(Class = No|Temperature = cool) = -\frac{3}{4} \times \log_2 \frac{3}{4} - \frac{1}{4} \times \log_2 \frac{1}{4}$

According to Table 7.4, we can derive the conditional entropy value for each of the two values of attribute 'Humidity'.

$E(Humidity = high) = -P(Class = Yes|Humidity = high) \times \log_2$
$P(Class = Yes|Humidity = high) - P(Class = No|Humidity = high) \times \log_2 P(Class = No|Humidity = high) = -\frac{3}{7} \times \log_2 \frac{3}{7} - \frac{4}{7} \times \log_2 \frac{4}{7}$

$E(Humidity = normal) = -P(Class = Yes|Humidity = normal) \times \log_2$
$P(Class = Yes|Humidity = normal) - P(Class = No|Humidity = normal)$
$\times \log_2 P(Class = No|Humidity = normal) = -\frac{1}{7} \times \log_2 \frac{1}{7} - \frac{6}{7} \times \log_2 \frac{6}{7}$

According to Table 7.5, we can derive the conditional entropy value for each of the two values of attribute 'Windy'.

$E(Windy = true) = -P(Class = Yes|Windy = true) \times \log_2 P(Class = Yes|Windy = true) - P(Class = No|Windy = true) \times \log_2 P(Class = No|Windy = true) = -\frac{1}{2} \times \log_2 \frac{1}{2} - \frac{1}{2} \times \log_2 \frac{1}{2}$

$E(Windy = false) = -P(Class = Yes|Windy = false) \times \log_2 P(Class = Yes|Windy = false) - P(Class = No|Windy = false) \times \log_2 P(Class = No|Windy = false) = -\frac{3}{4} \times \log_2 \frac{3}{4} - \frac{1}{4} \times \log_2 \frac{1}{4}$

According to the above illustration, the attribute-value pair $Outlook = overcast$ is chosen, since the conditional entropy value of it is the minimum (0). Also, the conditional entropy value of 0 indicates that there is no uncertainty any more, so the learning of the first rule is complete and the rule is expressed as: if $Outlook = overcast$ then $class = yes$. Following the completion of learning the first rule, all four instances with the attribute-value pair $Outlook = overcast$ are deleted from the training set, and the learning of the second rule is started on the reduced training set.

Existing methods typically involve selecting an attribute-value pair iteratively based on a single heuristic employed. For example, the IEBRG algorithm illustrated above involves selecting attribute-value pairs based on conditional entropy. However, as argued in [13], this strategy of attribute-value pair selection could lead to the learning of a rule set that has some rules of high quality but also others of low quality. In order to address this issue, we present a proposed multi-granularity framework for separate and conquer rule learning in Sect. 7.2.

7.2   Granular Framework of Learning

The purpose of proposing multi-granularity rule learning is to optimize the selection
of each attribute-value pair towards specializing the left hand side of a rule. The
proposed multi-granularity framework of rule learning is illustrated in Fig. 7.4.

This framework involves three levels of granularity as follows:

1. Level 1: The aim is at learning a rule set by using a heuristic and the evaluation
   of learning performance is through looking at the overall quality of a rule set.
2. Level 2: The aim is at learning each single rule by combining different heuristics
   in a competitive way, i.e. each heuristic is used for iterative selection of attribute-
   value pairs towards learning a rule and the rule of the best quality is added into
   the rule set. In this context, the evaluation of learning performance is through
   looking at the quality of each single rule.
3. Level 3: The aim is at learning each term of a rule by combining different heuristics
   in a competitive way, i.e. each heuristic is used for selection of an attribute-value
   pair as a rule term and the attribute-value pair that tends to optimize the quality
   of the rule is selected as a term (added into the left hand side of the rule). In this
   context, the evaluation of learning performance is through looking at the impact
   of each rule term towards optimizing the rule quality.

In the setting of granular computing, this framework involves explicitly both
granulation and organization. In particular, granulation is to separate a subset of
training instances towards selection of attribute-value pairs and learning of a rule.
Organization is to integrate rules (learned at different iterations) into a rule set.
According to Fig. 7.4, all selected rules in level 2 make up a rule set and all selected
rule terms in level 3 make up each single rule to be added into a rule set.

In terms of level 2 of the multi-granularity framework, a collaborative rule learning
approach has been proposed in [13] and is illustrated in Fig. 7.5. This approach
involves employing several heuristics such as entropy and probability for attribute-
value pair selection. Following the general procedure of separate and conquer rule
learning, at each interation, several rules are learned by using different heuristics,

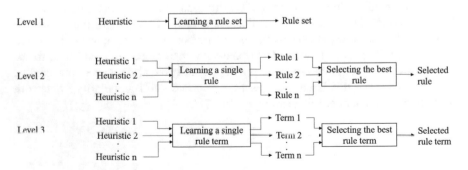

Fig. 7.4 Multi-granularity framework of rule learning

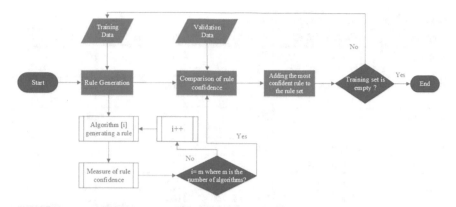

Fig. 7.5 Procedure of collaborative rule learning

and these rules are then compared in terms of their quality (some measures of rule quality can be found in [11]). Following the comparison of these rules, only the rule of the best quality is selected and added into the rule set.

7.3   Discussion

An experimental study has been conducted in [13] by using 20 UCI data sets and the characteristics of the data sets are shown in Table 7.6. Also, the heuristics employed for the Prism and IEBRG algorithms (conditional probability and conditional entropy) are combined for collaborative rule learning. In addition, confidence, J-measure, lift and leverage are used, respectively, for evaluation of rule quality. The results are shown in Table 7.7.

The results presented in Table 7.7 shows that the collaborative rule learning approach outperforms both Prism and IEBRG in 18 out of 20 cases, while the quality of each rule is evaluated by using confidence, J-measure, lift and leverage (shown in column 3-7 of Table 7.7. This indicates that the combination of different heuristics usually leads to advances in overall accuracy of classification.

In some cases (on kr-vs-kp, segment and credit-a data sets), the collaborative rule learning approach even outperforms both Prism and IEBRG to a large extent. This phenomenon can support the argument that different heuristics can be complementary to each other, especially on the basis that they have different advantages and disadvantages and that they are combined in an effective way.

On the other hand, the results show that this collaborative approach has a bias on the chosen measure of rule quality. It can be seen from Table 7.7 on the data sets anneal, ionosphere, iris, car, breast-w and mushroom, that at least one of the measures of rule quality fails to help outperform both Prism and IEBRG. This phenomenon could be due partially to the variance resulting from random partitioning of data, but

Table 7.6 Data sets for collaborative rule learning [13]

Dataset	Feature Types	#Attributes	#Instances	#Classes
anneal	discrete, continuous	38	798	6
credit-g	discrete, continuous	20	1000	2
diabetes	discrete, continuous	20	768	2
heart-stalog	continuous	13	270	2
ionosphere	continuous	34	351	2
iris	continuous	4	150	3
kr-vs-kp	discrete	36	3196	2
lymph	discrete, continuous	19	148	4
segment	continuous	19	2310	7
zoo	discrete, continuous	18	101	7
wine	continuous	13	178	3
breast-cancer	discrete	9	286	2
car	discrete	6	1728	4
breast-w	continuous	10	699	2
credit-a	discrete, continuous	15	690	2
heart-c	discrete, continuous	76	920	4
heart-h	discrete, continuous	76	920	4
hepatitis	discrete, continuous	20	155	2
mushroom	discrete	22	8124	2
vote	discrete	16	435	2

still indicates that it is critical to appropriately choose the measure of rule quality to reduce the bias resulting from the setting of rule quality evaluation.

Overall, the results indicate that the collaborative rule learning approach is useful for improving the quality of each single rule learned and thus improving the overall accuracy. In machine learning tasks, the main concern of a rule learning algorithm is typically about using a rule set as a whole to classify accurately unseen instances. In this context, some rules of low quality may be rarely or even never used for classification. In this case, although the accuracy may not be seriously affected, the improvement for the quality of each single rule is still necessary towards the improvement of overall accuracy, especially when a large set of test instances are used.

Table 7.7 Accuracy performed by collaborative rule learning [13]

Dataset	Prism	IEBRG	CRG(Conf)	CRG(J-measure)	CRG(Lift)	CRG (Leverage)
anneal	0.68	0.90	0.90	0.90	0.91	0.91
credit-g	0.69	0.67	0.72	0.72	0.71	0.72
diabetes	0.71	0.70	0.73	0.73	0.72	0.75
heart-Statlog	0.64	0.66	0.71	0.74	0.75	0.74
ionosphere	0.87	0.81	0.84	0.86	0.86	0.88
iris	0.72	0.93	0.92	0.94	0.95	0.95
kr-vs-kp	0.77	0.83	0.95	0.93	0.92	0.92
lymph	0.68	0.70	0.75	0.75	0.75	0.78
segment	0.55	0.68	0.80	0.80	0.81	0.77
zoo	0.62	0.79	0.87	0.85	0.84	0.88
wine	0.80	0.91	0.92	0.94	0.94	0.95
breast-cancer	0.67	0.69	0.71	0.71	0.73	0.72
car	0.71	0.76	0.76	0.75	0.77	0.76
breast-w	0.90	0.95	0.92	0.94	0.95	0.93
credit-a	0.69	0.69	0.77	0.79	0.80	0.80
heart-c	0.70	0.69	0.76	0.74	0.76	0.75
heart-h	0.76	0.74	0.83	0.81	0.82	0.83
hepatitis	0.82	0.82	0.83	0.85	0.84	0.86
mushroom	0.97	0.98	0.97	0.97	0.97	0.98
vote	0.88	0.90	0.93	0.96	0.95	0.95

Furthermore, rule learning can still be advanced further by extending the above collaborative approach. In particular, the way of using each heuristic for learning a single rule may lead to the case that some of the selected attribute-value pairs may impact positively on the rule quality, but some others may impact negatively on the quality. In other words, in the process of learning a rule, the quality of the rule may be improved at some iterations but also be reduced at other iterations. This indicates the necessity of investigating the way of combining different heuristics illustrated in level 3 of the proposed multi-granularity framework, towards advancing further the performance of rule learning.

References

1. Liu, H., A. Gegov, and M. Cocea. 2016. *Rule based systems for big data: a machine learning approach*. Switzerland: Springer.
2. H. Liu, A. Gegov, and F. Stahl. 2014. Categorization and construction of rule based systems. In *15th International Conference on Engineering Applications of Neural Networks*, Sofia, Bulgaria, 5–7 September 2014, 183–194.
3. Liu, H., and A. Gegov. 2015. *Collaborative decision making by ensemble rule based classification systems*. Switzerland: Springer.
4. R. J. Quinlan. 1993. *C4.5: programs for machine learning*. San Francisco: Morgan Kaufmann Publishers.
5. Furnkranz, J. 1999. Separate-and-conquer rule learning. *Artificial Intelligence Review* 13: 3–54.
6. Quinlan, R.J. 1986. Induction of decision trees. *Machine Learning* 1 (1): 81–106. March.
7. Breiman, L., J. Friedman, C.J. Stone, and R. Olshen. 1984. *Classification and Regression Trees*. Monterey, CA: Chapman and Hall/CRC.
8. Cendrowska, J. 1987. Prism: An algorithm for inducing modular rules. *International Journal of Man-Machine Studies* 27: 349–370. May.
9. Liu, H., and A. Gegov. 2016. *Induction of modular classification rules by information entropy based rule generation*. Switzerland: Springer.
10. Kononenko, I., and M. Kukar. 2007. *Machine learning and data mining: Introduction to principles and algorithms*. Chichester: Horwood Publishing Limited.
11. Liu, H., A. Gegov, and M. Cocea. 2016. Rule based systems: A granular computing perspective. *Granular Computing* 1 (4): 259–274.
12. H. Liu, M. Cocea, and A. Gegov, "Interpretability of computational models for sentiment analysis. In *Sentiment Analysis and Ontology Engineering: An Environment of Computational Intelligence*, vol. 639, eds. W. Pedrycz and S.-M. Chen, 199–220.
13. Liu, H., A. Gegov, and M. Cocea. 2016. Collaborative rule generation: An ensemble learning approach. *Journal of Intelligent and Fuzzy Systems* 30 (4): 2277–2287.

Chapter 8
Case Studies

Abstract In this chapter, we present several case studies on biomedical data processing and sentiment analysis. Biomedical data processing involves measuring of veracity and variability, respectively. In the sentiment analysis case study, we show the performance of fuzzy approaches on movie reviews data, in comparison with other commonly used non-fuzzy approaches.

8.1 Biomedical Data Processing

In this section, we present two case studies on biomedical data processing, based on the results reported in [1]. The first one addresses the veracity aspect, and is designed to show that cross-validation can be used to measure the learnability of algorithms on particular training data towards effective employment of learning algorithms. The second case study addresses variability, and is designed to show how the variability of data can be measured through checking the variance of the performance of a particular algorithm, when learning from the same data in independently repeated experiments.

The first case study was conducted by using 10 data sets retrieved from the Biomedical data repository [2]. The characteristics of these data sets are described in Table 8.1.

In particular, all these selected data sets are of high dimensionality and have additional test sets supplied. This selection is in order to support the experimental setup, which employs cross-validation towards measuring the learnability of particular algorithms on the training data and then employ suitable ones to build models that are evaluated by using test instances. In other words, for each of the selected data sets, the whole training set is provided in order to measure the extent to which a particular algorithm is suitable to build a model on the training set, and the test set is used to evaluate the performance of the model learned by using the algorithm.

In this setup, the results would show the extent to which the learnability of an algorithm measured by using cross validation on training data can be used as a good basis for judging whether the algorithm can lead to producing a confident model that performs well on additional test data. In this case study, C4.5, Naïve Bayes (NB)

© Springer International Publishing AG 2018

H. Liu and M. Cocea, *Granular Computing Based Machine Learning*,
Studies in Big Data 35, https://doi.org/10.1007/978-3-319-70058-8_8

Table 8.1   Bio-medical Data sets [1]

Name	Attribute types	Attributes	Instances	Classes
ALL-AML	Continuous	7130	72	2
DLBCL-NIH	Continuous	7400	160	2
lungCancer	Continuous	12534	32	2
MLL-Leukemia	Continuous	12583	72	3
BCR-ABL	Continuous	12559	327	2
E2A-PBX1	Continuous	12559	327	2
Hyperdip50	Continuous	12559	327	2
MLL	Continuous	12559	327	2
T-ALL	Continuous	12559	327	2
TEL-AML1	Continuous	12559	327	2

Table 8.2   Learnability on training data and prediction accuracy on test data [1]

Dataset	C4.5 I (%)	C4.5 II (%)	NB I (%)	NB II (%)	KNN I (%)	KNN II (%)
ALL-AML	93	100	70	71	88	97
DLBCL-NIH	44	58	55	63	56	63
LungCancer	94	89	25	90	88	97
MLL-Leukemia	79	100	22	53	89	100
BCR-ABL	91	95	96	95	97	96
E2A-PBX1	96	87	92	95	98	88
Hyperdip50	91	88	81	80	94	98
MLL	94	97	94	95	97	100
T-ALL	91	100	87	87	55	99
TEL-AML1	95	95	76	76	98	98

and K nearest neighbors (KNN) are chosen as learning algorithms for testing due to their popularity in real applications.

The results are shown in Table 8.2. In particular, C4.5 I means testing the learnability of the algorithm by cross validation on the basis of training data and C4.5 II means testing the performance of the predictive model using the additionally supplied test data. The same also applies to NB and KNN.

Table 8.2 shows that in almost all cases the learnability of an algorithm measured by using cross validation is effective for judging the suitability of an algorithm for a particular training set, which leads to expected performance on the corresponding test set. In other words, the results show that if an algorithm is judged to be suitable for a particular training set through measuring the learnability of this algorithm, then the model learned by the algorithm from the training set usually performs well on the additionally supplied test set.

On the other hand, when an algorithm is judged to be unsuitable for a particular training set through cross-validation, the results generally indicate the phenomenon that the model learned by the algorithm from the training set performs a low level of classification accuracy on the additionally supplied test set. In particular, it can be seen on the DLBCL-NIH data that all these three algorithms are judged to be less suitable for the training set and the models learned by these algorithms from the training set fail to perform well on the corresponding test set. Another similar case can be seen on the MLL-Leukemia data that NB is judged to be unsuitable for the training set and the model learned by the algorithm fails to perform well on the corresponding test set.

In addition, there are two exceptional cases on the lung-cancer and T-All data. In the first case, NB is judged to be very unsuitable for the training set but the performance on the test set by the model learned by the algorithm from the training set is very good. In the second case, KNN is judged to be less suitable for the training set but the actual performance on the test set by the model learned by the algorithm from the training set is extremely good. For both cases, it could be because the training set essentially covers the complete information and the split of the training set for the purpose of cross validation could result in incompleteness to which both NB and KNN are quite sensitive. However, when the algorithm learns from the whole training set, the resulted model covers the complete information from the training set and thus performs well on the test set.

The second case study was conducted by using 10 data sets retrieved from the UCI [3] and the biomedical repositories. The characteristics of the selected data sets are shown in Table 8.3.

The data sets selected from the UCI repository are all considered as small data as they are of lower dimensionality and sample size. On the other hand, the last five

Table 8.3 Data sets from UCI and biomedical repositories [1]

Name	Attribute types	Attributes	Instances	Classes
Diabetes	Discrete, continuous	20	768	2
Heart-statlog	Continuous	13	270	2
Hepatitis	Discrete, continuous	20	155	2
Ionosphere	Continuous	34	351	2
lymph	discrete, continuous	19	148	4
ColonTumor	Continuous	2001	62	2
DLBCLOutcome	Continuous	7130	58	2
DLBCLTumor	Continuous	7130	77	2
DLBCL-Stanford	Continuous	4027	47	2
Lung-Michigan	Continuous	7130	96	2

data sets selected from the biomedical repository are all considered as big data due to the fact that they are of high dimensionality. This selection is in order to put the case study in the context of data science by means of processing data with different scalability. In addition, all these chosen data sets are not supplied additional test sets. The selection of the data sets was also made so that both discrete and continuous attributes are present, which is in order to investigate how the different types of attributes could impact on the data variability.

On the basis of the chosen data, the experiment on each data set is undertaken by independently repeating the training-testing process 100 times and checking the variance of the performance over the 100 repetitions, on the basis of random sampling of training and test data in the ratio of 70:30. This experimental setup is in order to measure the extent to which the data is variable, leading to variance in terms of performance in machine learning tasks. In this context, C4.5, NB and KNN are chosen as learning algorithms for testing the variance due to the fact that these algorithms are not stable, i.e. they are sensitive to the changes in data sample. The results are shown in Table 8.3.

It can be seen from Table 8.4 that on each data set, while different algorithms are used, the standard deviation of the classification accuracy over 100 independently repeated experiments appears to be in a very similar level, except for the DLBCL-Standford data set on which NB displays a much lower level of standard deviation.

On the other hand, while looking at different data sets, the standard deviation for them appears to be very different no matter which one of the three algorithms is adopted. In particular, for the 5 UCI data sets, the standard deviation is lower than 5% in most cases or a bit higher than 5% in several cases (e.g. on the lymph and hepatitis data sets). In contrast, for the last five data sets selected from the biomedical repository, the standard deviation is usually higher than 5% and is even close to or higher than 10% in some cases (e.g. on the colonTumer and DLBCLOutcome data sets). An exceptional case happens from the lung-Michigan data set, which appears

Table 8.4 Data variability measured by standard deviation of classification accuracy [1]

Dataset	C4.5	NB	KNN
Diabetes	0.027	0.028	0.027
Heart-statlog	0.044	0.039	0.035
Hepatitis	0.046	0.042	0.073
Ionosphere	0.031	0.043	0.035
lymph	0.057	0.057	0.055
ColonTumor	0.094	0.105	0.089
DLBCLOutcome	0.122	0.104	0.109
DLBCLTumor	0.074	0.067	0.072
DLBCL-Stanford	0.133	0.060	0.096
Lung-Michigan	0.040	0.041	0.028

to have the standard deviation lower than 5%, no matter which one of the three algorithms is used.

In addition, it can also be seen from Table 8.4 that the data sets that contain only continuous attributes appear to have the standard deviation higher than the data sets that contain discrete attributes. In fact, the presence of continuous attributes generally increases the attribute complexity, and thus makes the data more complex, which leads to the potential increase of the data variability.

The results shown in Table 8.4 generally indicate that attribute complexity, data dimensionality and sample size impact on the size of data and that data with a larger size is likely to be of higher variability, leading to higher variance in terms of performance in machine learning tasks, especially when the training and test data are sampled on a purely random basis.

8.2 Sentiment Analysis

In this section, we present a case study on sentiment analysis, based on the results reported in [4]. This experimental study was conducted by using four polarity data sets on movie reviews. The data sets with the number of instances in the positive and negative categories are listed in Table 8.5 and more details can be found in [5–7].

All the experiments were conducted using the following procedure:

- Step 1: The textual data is enriched by using the Part-of-Speech (POS) Tagger and the Abner Tagger [8];
- Step 2: The enriched data is transformed through using the Bag-of-Words (BOW) method;
- Step 3: For each word, its relative frequency, absolute frequency, inverse category frequency and inverse document frequency are calculated towards filtering out words with low frequency;
- Step 4: The words left following the last stage are preprocessed by filtering stop words, words with no more than N Chars and numbers, stemming porter and erasing punctuation;
- Step 5: Each document is turned into a vector that consists of all the words appearing in the textual data set, each of which is turned into a numerical attribute that reflects the frequency of the word;

Table 8.5 Data sets on movie review with number of positive and negative instances [4]

Data set	Positive	Negative
PolarityDataset V 0.9	700	700
PolarityDataset V 1.1	700	700
PolarityDataset V 1.0	700	700
PolarityDataset V 2.0	1000	1000

- Step 6: All the document vectors are classified to be positive or negative through using machine learning algorithms.

The BOW method mentioned in Step 2 generally means extracting terms (defined as different numbers of words, e.g. 1-word terms, 2-word terms, etc.) from the text and counting the frequency of each term. The most frequent approach is for a term to correspond to a single word. The following example is given for illustration:

Here are two text instances:

1. Alice encrypts a message and sends it to Bob.
2. Bob receives the message from Alice and decrypts it.

Based on the two instances above, a list of distinct words is created:
[Alice, Bob, encrypts, decrypts, sends, receives, message, a, the, and, it, from, to]
Two feature vectors for the two instances are created:

1. [1, 1, 1, 0, 1, 0, 1, 1, 0, 1, 1, 0, 1]
2. [1, 1, 0, 1, 0, 1, 1, 0, 1, 1, 1, 1, 0]

In the above two feature vectors, each numerical value represents the frequency of a corresponding word.

In the text classification stage, the structured data set is divided into a training set and a test set in the ratio of 70:30.

We show the performance of a fuzzy rule based approach in terms of classification accuracy, in comparison with the commonly used learning approaches, i.e. Naive Bayes (NB) and C4.5, which are known to perform well for sentiment analysis.

The results are shown in Table 8.6 and indicate that the fuzzy approach performs slightly better than the well known NB and C4.5 methods, thus indicating the suitability of fuzzy approaches for sentiment analysis tasks.

On the other hand, as reported in [9], the results presented in Table 8.7 show empirically that sentiment data is generally of massively high dimensionality following feature extraction by transforming textual data into structural data through use of the BOW method. Even after any irrelevant words have been filtered, the data dimensionality (the number of words) is still very high (over thousands). The experimental results could generally indicate that the interpretation of computational models is much constrained and interpretability is thus an issue that is needed to be

Table 8.6 Accuracy of sentiment classification [4]

Data set	Naive bayes	C4.5	Fuzzy rules
PolarityDatasetV0.9	0.936	0.942	0.951
PolarityDatasetV1.1	0.942	0.945	0.951
PolarityDatasetV1.0	0.939	0.939	0.962
PolarityDatasetV2.0	0.913	0.938	0.943

Table 8.7 Number of words extracted through using BOW and number of words left after filtering low frequent words [4]

Data set	#words	#words(left)
PolarityDataset V 0.9	523456	1014
PolarityDataset V 1.1	515503	1027
PolarityDataset V 1.0	517567	1030
PolarityDataset V 2.0	726250	1030

dealt with through addressing the data dimensionality issue, for which the use of fuzzy information granulation was proposed in [9].

As introduced in Chap. 1, granulation is generally aimed at decomposing a large granule (in a higher level of granularity) into several smaller granules (in a lower level of granularity). According to different formalisms of information granulation, the corresponding granules are of different types, such as crisp granules, probabilistic granules, fuzzy granules and rough granules.

In the context of text processing, information granules are typically of fuzzy type, such as sections, subsections, paragraphs, passages, sentences, phrases and words. The above examples of fuzzy information granules are actually in different levels of granularity so we propose a multi-granularity approach of text processing in this section. In particular, textual data is decomposed into several parts and each of these parts may be divided again depending on its complexity, through fuzzy information granulation.

As mentioned earlier in this section, feature extraction from textual data usually results in massively high dimensionality, which leads to difficulty in the interpretation of computational models. This is mainly because the bag-of-words (BOW) method is used too early for transforming textual data into structural data. In other words, traditional approaches of text processing only involve single-granularity learning, and all features extracted through using the BOW method are global ones. In fact, an instance of textual data can be decomposed into sub-instances in the setting of granular computing. In this context, text processing can involve multi-granularity learning and there could be more local features extracted from those sub-instances of the original textual instances. For example, text can be divided into phrases, and a document can be decomposed into several sections, each of which can be again divided into subsections. Therefore, information granules in different levels of granularity would involve different local features to be extracted. The above way of text processing is also in line with the main requirements of big data processing, namely, decomposition, parallelism, modularity and recurrence [10], which can lead to the reduction of instance complexity so that each instance of textual data (as an information granule) can have its dimensionality and fuzziness reduced.

Overall, the above approach of text processing involves multi-granularity learning, which decomposes a textual data set into several modules/sub-modules so that each module/sub-module can be much less complex (of much lower dimensionality and fuzziness), and enables the extraction of local features from each module/sub-module of original textual data. In addition, the above approach also leads to the reduction of computational complexity, since parallelism can be involved in processing the modules/sub-modules of textual data following the decomposition of the data.

In order to adopt the above multi-granularity approach of text processing, there are four questions that need to be considered as follows:

1. How many levels of granularity are required?
2. Is text clustering required towards the reduction of data size through modularizing a textual data set?
3. In each level of granularity, how many information granules are involved?
4. At which level of granularity should the BOW method be used for transforming textual data into structural data?

With regard to question 1, the number of granularity levels partially depends on the type of text. In other words, text can be of different scalability, such as documents, comments and messages. Documents usually do not have any word limits, and thus can be very long and complex resulting in massive dimensionality, if information granulation is not adopted. However, documents are generally well structured leading to a more straightforward way of information granulation based on different levels of headings, e.g. sections and subsections. In addition, paragraphs in each section/subsection generally still need to be divided further into passages/sentences towards reaching the bottom level of granularity for words, which indicates that the number of granularity levels is generally greater than the number of heading levels in a text document.

Comments are typically involved on any web platforms such as social media, forums and e-learning environments. In this context, comments are usually limited to a small number of words, e.g. 200 words. Therefore, the dimensionality issue mentioned above is less likely to arise when comparing with document processing. However, comments are typically not structured, which results in the difficulty in information granulation. In this case, the number of granularity levels depends highly on the complexity of text, i.e. the top level of granularity may be paragraphs or passages, while the bottom level is typically words.

Messages are also typically involved on web platforms, but the number of words is generally limited to a few words/sentences, unlike comments. Therefore, the issue on massive dimensionality is much less likely to arise, but messages, similar to comments, are not well structured, which also results in the difficulty in information granulation. In this case, the number of granularity levels also depends highly on the complexity of text, i.e. the top level of granularity may be sentences or phrases with the bottom level consisting typically of words.

With regard to question 2, text clustering is needed typically in two cases. Firstly, when the training data is large, it is very likely to involve a large total number of words resulting in the massive dimensionality problem. In addition, large training data is also likely to contain instances in different contexts, which makes a learning task less focused and thus shallow. Secondly, when the textual data is in the form of documents, each document would usually contain much more words than a comment or a message, which is still more likely to result in the massive dimensionality problem. Therefore, in the above two cases, text clustering is highly required towards the reduction of data dimensionality and having more focused learning in depth.

With regard to question 3, the number of information granules involved in each level of granularity depends on the consistency of structure among instances of textual data. For example, a training set of documents can be of the exactly same structure or different structures. In the former case, information granulation for each of the documents in a particular level of granularity is simply undertaken based on the document headings in the corresponding level, e.g. information granulation in level one is simply done by having each heading 1 with its text contents as an information granule in this level of granularity. In the latter case, the number of information granules needs to be determined based upon the structure complexity of the documents on average. This is very similar to the problem of determining the number of clusters on the basis of the given training instances. In this context, each information granule can be interpreted as a deterministic/fuzzy cluster of training instances of high similarity. For textual data, each information granule would represent a cluster of sub-instances of textual training instances.

With regard to question 4, it is highly expected that the BOW method is not adopted until each information granule in a particular level of granularity is small and simple enough. In this case, the dimensionality of training data from each information granule (cluster) is much reduced comparing with traditional approaches of text processing, which involve direct use of BOW on the basis of original textual data. For example, a section may have a number of sub-sections. In this context, the first paragraph is generally aimed at outlining the whole section, which is typically short and simple, so BOW can be used immediately at this point for transforming the text of this paragraph or it is used shortly following a simple decomposition of this paragraph. However, for all the other paragraphs in this section that directly belong to its subsections, it is not expected to adopt BOW immediately at this point, since these paragraphs still need to be moved into other granules located in the next deeper level of granularity.

The multi-granularity approach of text processing is illustrated in Fig. 8.1 This illustration is based on the following scenario: each document is a research paper, which consists of four main sections, i.e. Sect. 1.4. Also, Sect. 3 contains two subsections, i.e. Sects. 3.1 and 3.2. In addition, an abstract is included as an independent part of text in each research paper.

Figure 8.1 indicates that the parent of an information granule may not necessarily be located in a direct upper level of granularity. For example, an abstract is an information granule that belong to the granularity of paragraphs but the parent of the information granule (abstract) is located in the top level of granularity (paper). Also,

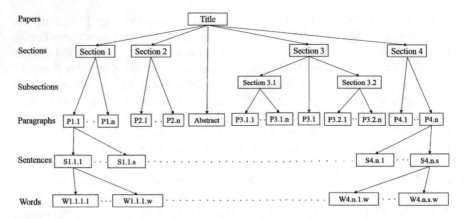

Fig. 8.1 Fuzzy information granulation for text processing [9]

a section may consist of several subsections but the first paragraph in this section typically directly belongs to this section rather than any subsections.

On the other hand, it is a normal phenomenon that the number of paragraphs involved in each section, especially for different documents (papers), is not deterministic. Therefore, information granulation in the level of granularity for paragraphs would be considered as a fuzzy granulation problem, since it is not deterministic to decide the number of information granules (paragraphs) provided from each section/subsection. In practice, it is even very likely to have different documents with different structures. From this point of view, the decision on the number of information granules in the level of granularity for sections/subsections is not deterministic either and thus it is also considered as a fuzzy granulation problem. On the basis of the above descriptions, in the last two levels of granularity for sentences and words (see Fig. 8.1), respectively, the information granulation also needs to be undertaken through fuzzy approaches, in terms of deciding the number of bags of sentences/words (BOS/BOW).

As mentioned in Chap. 1, granular computing involves both granulation and organization. In general, the former is a top-down process and the latter is a bottom-up process. Decomposition of a text document into smaller granules belongs to granulation. Following this granulation, organization is required to get the final classification for test instances, i.e. documents. In this context, as shown in Fig. 8.1, there are a number of granules in each level of granularity, and each of the information granules is typically interpreted as a fuzzy cluster. In the testing stage, each test instance (document) is divided recursively into sub-instances which are located in different levels of granularity. In each level of granularity, each sub-instance is related to several particular information granules, depending if the parents of the particular information granules relate to the parent of the sub-instance, and each sub-instance is also assigned a certain degree of fuzzy membership to each of the related information granules (fuzzy clusters), following the fuzzification step illustrated in Chap. 5.

Furthermore, each sub-instance is inferred by these related fuzzy information granules towards finalising the fuzzy membership degree of the sub-instance to each of the given classes (e.g. positive and negative), following the inference step (that consists of application, implication and aggregation) as illustrated in Chap. 5.

Furthermore, each sub-instance is inferred by these related fuzzy information granules towards finalising the fuzzy membership degree of the sub-instance to each of the given classes (e.g. positive and negative), following the inference step (that consists of application, implication and aggregation) as illustrated in Chap. 5. Finally, the fuzzy membership degrees of these sub-instances (to all of the given classes) need to be aggregated through disjunction towards providing an overall degree of fuzzy membership (to each of the classes) for the parent of these sub-instances. For example, a sentence S has two sub-instances W_1 and W_2 located in a lower level of granularity and S belongs to one of the two classes: positive and negative. In this case, if the fuzzy membership degrees of W_1 to the positive and negative classes are 0.7 and 0.3, respectively, and the degrees of W_2 to the two classes are 0.5 and 0.5, respectively, then the fuzzy membership degrees of S to the two classes are 0.7 and 0.5, respectively.

On the basis of the above paragraph, except for the top and bottom levels of granularity, each of sub-instances in a particular level would be given two sets of fuzzy membership degrees. In particular, one of the set of fuzzy membership degrees is provided from disjunction of the fuzzy membership degrees of the sub-subinstances of a particular sub-instance and the other set of the membership degrees is provided from the inference by the related granules (fuzzy clusters) in this level of granularity. However, the appearance of the two sets of fuzzy membership degrees raises the question: how are the two sets of fuzzy membership degrees combined towards having an overall set of fuzzy membership degrees for each sub-instance in each level of granularity (except for the top and bottom levels)? This research direction is further discussed in Chap. 9.

In terms of the interpretation of prediction results, from each level of information granularity, the fuzzy membership degrees of each sub-instance to all of the given classes are shown explicitly. Also, the hierarchical relationships between a sub-instance and each of its sub-subinstances can be shown clearly. Therefore, the final result of classifying a test instance can be derived implicitly through the bottom up process as described in the above two paragraphs. This derivation can also be described in natural language to facilitate interpretability; for examples, an output at paragraph level could be expresses as "this paragraph contains 3 positive sentences and 2 negative sentences". In addition, the fuzzy membership degrees can also be given as an output for each of the sentences in the paragraphs based on which the above output was created.

References

1. Liu, H., A. Gegov, and M. Cocea. 2017. Unified framework for control of machine learning tasks towards effective and efficient processing of big data. In *Data Science and Big Data: An Environment of Computational Intelligence*, 123–140. Switzerland: Springer.
2. J. Li and H. Liu. 2003. Kent ridge bio-medical dataset. http://datam.i2r.a-star.edu.sg/datasets/krbd/.
3. M. Lichman. 2013. UCI machine learning repository. http://archive.ics.uci.edu/ml.
4. H. Liu and M. Cocea. 2017. Fuzzy rule based systems for interpretable sentiment analysis. In *International Conference on Advanced Computational Intelligence*, Doha, Qatar, 4–6 February 2017, 129–136.
5. Pang, B., L. Lee, and S. Vaithyanathan. 2002. Thumbs up? sentiment classification using machine learning techniques. *Proceedings of EMNLP* 2002: 79–86.
6. B. Pang and L. Lee. 2004. A sentimental education: Sentiment analysis using subjectivity. In *Proceedings of ACL*, 2004, 271–278.
7. B. Pang and L. Lee. 2005. Seeing stars: Exploiting class relationships for sentiment categorization with respect to rating scales. In *Proceedings of ACL*, 115–124.
8. Thiel, K., and M. Berthold. 2012. *The knime text processing feature: An introduction*. KNIME: Technical Report.
9. H. Liu and M. Cocea. Fuzzy information granulation towards interpretable sentiment analysis. *Granular Computing*, 3 (1), In press.
10. Wang, L., and C.A. Alexander. 2016. Machine learning in big data. *International Journal of Mathematical, Engineering and Management Sciences* 1 (2): 52–61.

Chapter 9
Conclusion

Abstract In this chapter, we stress the contributions and importance of this book from both scientific and philosophical perspectives. In particular, we describe the theoretical significance, practical importance and methodological impacts of our work presented in this book. We also show how the proposal of granular computing based machine learning is inspired philosophically from real-life examples. Moreover, we suggest some further directions to extend the current research towards advancing machine learning in the future.

9.1 Scientific Aspects

The focus of this book was to introduce our proposed approaches of granular computing based machine learning towards the shift from shallow learning to deep learning (in its broader sense). In particular, the following transformations have been proposed: (a) supervised-learning to semi-supervised learning, (b) heuristic learning to semi-heuristic learning, (c) single-task learning to multi-task learning, (d) discriminative learning to generative learning (e) random data partitioning to semi-random data partitioning. All these transformations have theoretical significance, practical importance and methodological impact.

In the context of semi-supervised learning, the proposal of machine based labelling presented in Chap. 3 shows novel applications of supervised learning algorithms. In particular, we argued in Chap. 3 that supervised learning algorithms need to be capable of learning from only a small part of the training set, i.e. it is not necessary for the algorithm to go through the whole training set towards learning a classifier. This argumentation is made based on the fact that incorrect labelling of instances would lead to negative impacts on learning from a training set partially labelled by machines. Since incorrect labelling must be avoided and the number of labelled training instances is very small, it is highly desirable to have learning algorithms that are capable of learning from a very small sample of training data. From this point of view, we propose the use of K Nearest Neighbours (KNN) [1] or Support Vector Machine (SVM) [2] for machine based labelling, due to the fact that both algorithms only need to take a small number of training instances towards building classifiers.

© Springer International Publishing AG 2018

H. Liu and M. Cocea, *Granular Computing Based Machine Learning*,
Studies in Big Data 35, https://doi.org/10.1007/978-3-319-70058-8_9

In practice, semi-supervised learning through machine based labelling can lead to reducing significantly the cost of manual labelling of data by experts, especially in the context of big data processing. Also, machine based labelling could help speed up the process of preparing data for learning tasks. In other words, once a small number of training instances have been labelled, then it would be ready to start the learning tasks, although the rest of the training instances still need to be labelled internally during the learning process. In addition, in laboratory research, machine based labelling also provides the space for comparing expert based labelling and machine based labelling in terms of their impacts on learning performance, when the test data is fully labelled by experts. In fact, expert based labelling could be subjective, especially in application areas such as sentiment analysis. In this case, machine based labelling may help verify the correctness of class labels assigned to training instances by experts, and identify highly ambiguous situations. In such situations, generative rather than discriminative learning may be more beneficial.

In the context of semi-heuristic learning, the proposal of probabilistic voting through nature inspiration presented in Chap. 4 shows novel applications of the theory of natural selection [3] as part of the procedure of Genetic Algorithm (GA) [4]. The proposal is aimed at reducing the bias towards the class of the highest frequency or weight, when voting is adopted for ensemble classification. In particular, we argued in Chap. 4 that following individual classifications from base classifiers, the class that is assigned the highest frequency or weight can only be considered to have the best chance of being the correct one assigned to the unseen instance, due to the incompleteness of training data.

On the basis of the above argumentation, we designed to assign this class the highest chance of being selected and assigned to the unseen instance, rather than to always assign this class to the unseen instance with no uncertainty. In this way, reduced bias can be achieved, but the variance in terms of classification performance may increase. The experimental results presented in Chap. 4 show that when probabilistic voting is adopted, the classification accuracy is usually improved, in comparison with the use of majority voting (based on frequency) or weighted voting (based on weight), i.e. the positive impact from reduction of bias tends to be higher than the negative impact from the increase in variance.

In practice, it is very likely that the training data collected represents only a sample, rather than a full population. In this context, it is very difficult to guarantee that complete knowledge can be learned from such a training set towards correctly classifying unseen instances. From this point of view, the case that a class has the highest frequency or weight only indicates the discovery from the current sample of training data by using an algorithm, but a different case may be discovered from an updated sample of training data, even if the same algorithm is used. The above argumentation again indicates the necessity of adopting probabilistic voting through nature inspiration.

On the other hand, in sentiment analysis, training data is typically collected through social media platforms such as Twitter. Due to the nature that the majority of users may tend to post more positive comments, it could happen that most of the collected training instances belong to the positive class through random data collection, which is likely to lead to the problem of class imbalance. For example, as mentioned in [5], the training data collected for cyberbullying detection only has 10% instances that belong to the negative class (cyberbullying). In this context, it is very likely to occur that classifiers learned from such training data are highly biased towards the positive class, which leads to the case that the positive class would always have the highest frequency or weight. The above case again indicates the necessity of adopting probabilistic voting through nature inspiration, towards reduction of bias in voting.

In the context of generative multi-task learning, we introduced novel applications of fuzzy logic in Chap. 5 for proving that different classes may not be mutually exclusive, i.e. there could be specific relationships between classes, such as mutual exclusion, correlation and mutual independence. In particular, we argued in Chap. 5 that if the sum of fuzzy membership degree values of an instance to different classes is higher than 1, then it would be an indication that these classes may involve other relationships such as correlation and mutual independence, especially when it appears to have the case that an instance has the membership degree value of 1 to more than one class. In contrast, if the sum of the membership degree values of an instance to two or more classes is always 1, then it is very likely to be the case that these two or more classes are mutually exclusive. In addition, fuzzy approaches can also be used to detect the case that an instance does not belong to any one of the predefined classes, when it appears to have the phenomenon that an instance has the membership degree value of 0 to all these classes.

In practice, there are many real-life problems that show the necessity of using generative multi-task learning through fuzzy approaches. For example, in the context of text classification, it is very likely that a document involves contents relating to different topics [6], so the document would be classified into different categories. In terms of relationships between classes, different topics could have correlations if these topics lie in similar areas, or these topics are independent of each other if they lie in dissimilar areas. The above argumentation indicates the necessity of generative multi-task learning. Since different topics are not mutually exclusive, it is more appropriate to undertake generative learning [7] instead of discriminative learning, i.e. these topics need to be treated equally in terms of judging the membership or non-membership of a document to each of these categories (topics). Also, multi-task learning needs to be undertaken instead of single-task learning, since the learning task is not to discriminate one topic from the other topics towards classifying a document into a category, but to judge if a document should be put into each specific category.

In comparison with traditional learning approaches, fuzzy approaches are more capable of dealing with fuzziness and imprecision, especially for text processing. In particular, traditional learning approaches treat classification as a black-white problem, i.e. an instance either belongs to a class or not. However, in reality, most problems are grey to some extent, i.e. there is not a clear boundary to separate different classes [8]. In this context, fuzzy approaches can be used to identify the degree to which an instance belongs to a class, i.e. there is a degree of fuzzy membership (ranged from 0 to 1). From social perspectives, sentiment analysis could be conducted towards identifying any social impacts. For example, in the context of cyberhate classification [9, 10], different hate speech may lead to different severity of anti-social issues. From this point of view, it is necessary not only to identify that the content of online speech is hateful, but also to measure the degree to which the content is hateful. The above argumentation indicates that fuzzy approaches are more suitable than traditional learning approaches, in terms of dealing with grey problems in real life.

On the other hand, fuzzy approaches are less sensitive to class imbalance [11, 12] in comparison with traditional learning approaches. As mentioned in Chap. 2, traditional learning approaches typically belong to discriminative learning, and thus tend to learn classifiers that are biased towards the majority class (with the highest frequency), if the training data is imbalanced. In contrast, as mentioned in Chap. 5, fuzzy approaches typically belong to generative learning, which treats all classes equally without discrimination between classes in the process of learning, so learning of classifiers is generally not affected by class imbalance of training data.

In the context of semi-random data partitioning, we introduced novel applications of granular computing concepts in terms of partitioning of granules [13]. In particular, we justified in Chap. 6 that random partitioning is undertaken inside a whole sample of data and semi-random partitioning is undertaken inside each of the sub-samples into which the whole sample is divided. In the setting of granular computing, we treat the whole sample of data as a granule and each of its sub-samples (instances of a specific class) as sub-granules. Also, we define that the whole sample is a granule located in level one (the top level) of granularity and that each sub-sample is a sub-granule located in level two (the next lower level) of granularity. In the above way of design, multi-granularity data partitioning can be achieved through sampling of training/test instances inside a sample/sub-sample located at a particular level of granularity.

In practice, the multi-granularity framework of data partitioning presented in Chap. 6 can be used to address two issues: (a) class imbalance and (b) sample representativeness, which could lead to the reduction of variance in learning performance. In particular, level 2 of the framework was designed to avoid class imbalance. We argued that the nature of random data partitioning may result in class imbalance even if the original sample of data is balanced. For example, if a data sample contains two classes ('positive' and 'negative') of instances in a 50/50 distribution, then random data partitioning may result in the case that more than 50% positive instances are selected as training instances and more than 50% negative instances are selected as test instances, i.e. the majority class is 'positive' in the training set but it is 'negative' in the test set. However, when the instances of different classes are separated through dividing the whole sample into sub-samples (each sub-sample contains instances of a specific class), class imbalance can be entirely avoided, since the class frequencies can be preserved through the way of semi-random partitioning designed at level two of the multi-granularity framework. A more detailed proof can be found in [14] and was also shown in Chap. 6.

9.2 Philosophical Aspects

This book focuses on scientific aspects; however, the ideas presented in this book also have some grounding in philosophical concepts, which are inspired from human learning and judgments, and are described as follows:

Supervised learning is philosophically similar to the strategy of learning with a teacher. For example, students are given revision questions alongside answers provided by teachers. In this context, students can fully identify if they have done all revision questions correctly, towards judging if they have grasped all relevant knowledge. Following the attempt on revision questions, students are given exam questions without answers, towards testing how well they have learned the relevant knowledge.

Semi-supervised learning is philosophically similar to the strategy of learning partially with a teacher and partially as self-learning. For example, students are given revision questions but only some of the questions are provided with answers. In this context, students need to first complete those questions that are provided with answers, and then judge if they have done correctly the other questions, based on those questions that are already given correct answers by teachers. Also, the presence of big data is philosophically similar to the case of having a large number of questions. From this point of view, it is not realistic to have all questions provided with answers by teachers, so semi-supervised learning is more encouraged in this case.

Heuristic learning is philosophically similar to the strategy of learning to make judgments based on principles introduced in textbooks or guidance notes. For example, in the context of recruitment, employers usually judge the capability of applicants through looking at their track record provided in their CV. In this context, the employers would normally decide to offer a post to the candidate who got the best qualifications and achievements in the past in comparison with the other candidates.

Semi-heuristic learning is philosophically similar to the strategy of learning to make judgments based partially on principles introduced in textbooks or guidance note, and partially on guess. For example, in the context of job applications, applicants would usually judge their suitability for a post based on their background and past achievements, i.e. they judge how well they have met all the essential criteria or even some of the desirable ones specified in the job descriptions. In this context, the applicants would normally send their applications for posts for which they have met all the essential criteria or even some desirable ones. Since they usually do not know the interests of employers nor do they have the knowledge about the background of other applicants, they can not say that they will definitely be successful in their applications, but only guess that they are very likely to be successful. In other words, applicants usually need to have construct their CV to maximise their chance of success, which is philosophically similar to increasing the positive bias in machine learning. However, due to the uncertainty about the interests of employers and the background of competitors, the application outcome could vary with some randomness, which is philosophically similar to the concept of variance in machine learning.

Single-task learning is philosophically similar to the case of learning to complete a single task. For example, students learn knowledge about a subject for passing an exam. In reality, a learning task could be very simple in a fundamental level of education (e.g. in primary schools or middle schools), so the strategy of single-task learning is appropriate.

However, a learning task could become more complex after students progress to a higher level of education (e.g. undergraduate or postgraduate). In this case, it is more necessary to turn single-task learning into multi-task learning, i.e. a task is divided into sub-tasks. For example, a module usually involves a number of learning outcomes, which indicates that students may need to set a task per learning outcome. Also, a module could involve multiple assessments, such as a coursework and an exam, or two or more courseworks, so students need to attempt all these assessments towards passing a module. In some cases, different assessments of a module could involve specific relationships, such as correlation and mutual independence, which is philosophically similar to the concept of relationships between granules in the context of granular computing.

Discriminative learning is philosophically similar to the task of learning to do single choice questions. For example, students need to select one of four given choices as the answer to a question. In this context, once a choice is identified to be the correct one as the answer to this question, it is a normal practice to stop analyzing the other choices, since these choices are mutually exclusive. Also, when it is uncertain to judge which choice is correct, it is a normal practice to compare these choices towards discriminating one choice from the other ones.

Generative learning is philosophically similar to the task of learning to do multiple choice questions. For example, students need to judge independently on each of four given choices whether they are correct. In this context, a multiple choice question is essentially done by transforming it into four judgment questions, i.e. the judgment on each choice does not affect the judgments on the other choices. In reality, a multiple choice question could mean that any combination of answers for each of the four choices is potentially the correct one. From this point of view, there may be two extreme cases despite their rare occurrence: (a) all the four choices are correct; (b) none of the four choices is correct. When the first case appears, it is a normal practice to provide all the choices together as the answer to this question. However, if the second case can truly come up, then it is needed to provide the fifth choice (none of the above is correct) to the question.

Random data partitioning into training and test sets is philosophically similar to random sampling of revision and exam questions. Also, the constraint that training and test sets must have no overlap is very similar to the rule that revision questions must be completely different from exam questions. In reality, random sampling of questions is very likely to result in the case that exam questions are highly dissimilar to revision questions, which leads to the outcome that students do badly in exams, although they have done well in revision. In other words, the testing does not cover what students have learned but mostly what they have not learned.

In fact, different questions would normally be related to different topics of learning. From this point of view, it is necessary that both revision and exam questions cover all the topics in a consistent percentage distribution. For example, there are four topics of maths, namely, mathematical logic, set theory, graph theory and linear algebra systems, and each of the four topics is weighted 25%. In this context, both revision and exam questions should cover all of the four topics in the percentage distribution: mathematical logic (25%), set theory (25%), graph theory (25%) and linear algebra systems (25%). In order to achieve the above requirement, it is necessary to adopt semi-random sampling of revision and exam questions, which is philosophically similar to semi-random partitioning of data in machine learning. Based on the above example, semi-random sampling is achieved through dividing the questions into four groups (each group corresponds to one of the four topics), and then sampling is done randomly within each group of questions, such that the percentage of questions for each topic is preserved.

In the context of granular computing, the concept of granules is very similar to many real-life examples, which involve forming groups of individual entities. In set theory, examples include deterministic sets, probabilistic sets, fuzzy sets and rough sets. In computer science, examples include classes, objects and clusters. In natural language processing (NLP), examples include documents, chapters, sections, subsections, paragraphs, sentences and words. In higher education, examples include universities, faculties, departments and research groups. In machine learning, examples include training sets, test sets and feature sets. In rule based systems, examples include rule bases, rule sets, rules and rule terms.

On the other hand, granules can be in different levels of granularity, which is very similar to many real-life examples. In set theory, a set can have subsets and each subset can also have sub-subsets, which leads to different levels of granularity. In computer science, a class can be specialized or decomposed into sub-classes. Similarly, a cluster can be specialized or decomposed into sub-clusters. Both of the above facts indicate the need to employ the concept of granularity. In NLP, textual contents can be organized in different levels of units, such as chapters and section, as mentioned above. Each of the these levels of units actually represents a level of granularity. In higher education, a university is normally organized to have different levels of units, such as faculties and departments, as mentioned above. Each of these unit levels also represents a level of granularity. In military bases, troops are organized to involve different levels of units, such as squad, platoon, company, battalion, regiment, division and corps. Each of these military levels can actually be viewed as a level of granularity.

Overall, granular computing is not only a scientific concept, but also a philosophical perspective, which provides people with a structured way of thinking. The two main operations, namely, granulation and organization, have been popularly involved in real life as the top-down and bottom-up approaches, respectively.

9.3 Further Directions

In this book, we have introduced several approaches of granular computing based machine learning, namely: (a) semi-supervised learning through machine based labelling, (b) nature inspired semi-heuristic learning, (c) fuzzy classification through generative multi-task learning, (d) multi-granularity semi-supervised data partitioning, and (e) multi-granularity rule learning. Also, we have introduced through a case study how to advance sentiment analysis by using fuzzy approaches. In order to advance this research area, we provide further directions as follows:

In terms of semi-supervised learning, we will investigate experimentally the performance of our proposed approaches (presented in Chap. 3) of machine based labelling through use of KNN and SVM. In particular, we will conduct experiments by randomly removing labels of a subset of training instances and then having the subset of training instances labelled through machine based labelling. However, test instances need to be kept the same as they are. In this way, we can compare the performance of learning through two different ways of labelling, i.e. one way leads to learning from expert labelled instances and the other way leads to learning from a set of instances partially labelled by experts/machines. If the two ways of labelling lead to similar performance of learning, this would be a strong indication that machine based labelling can be adopted with a high level of confidence, instead of expert based labelling.

In terms of semi-heuristic learning, we will extend the nature inspired framework of ensemble learning presented in Chap. 4 by making two modifications. The first modification is to employ multiple learning algorithms for learning a group of base classifiers from each sample of training data, and within each group of base classifiers, one of them is selected naturally for classifying test instances. The nature selection is based on the overall accuracy of these base classifiers evaluated by using validation data. The second modification is to employ precision (with respect to a class) for probabilistic voting, instead of overall accuracy. In this way, it is expected to reduce the bias in terms of both employing base classifiers and voting towards increasing the overall accuracy of classification.

In terms of fuzzy classification through generative multi-task learning, we will investigate experimentally the performance of generative learning of fuzzy rules through the multi-granularity framework presented in Chap. 5. Also, as mentioned in [6], multi-task classification can be achieved through both supervised learning and semi-supervised learning. In particular, for supervised learning, data labelling needs to be done by transforming the class attribute into several binary attributes, each of which is corresponding to a class label. In this way, experts need to judge on each class whether an instance belongs to it by assigning a truth value (0 or 1).

For semi-supervised learning, data sets, which have been used previously in traditional classification tasks, can be used again by transforming the class attribute into several binary attributes. In this way, the transformed data set would have all the binary attributes assigned truth values of 0 or 1. If an instance has one of its binary attributes assigned 0, this would mean that the instance has not been labelled on the corresponding class. Otherwise, the instance would have been labelled on the corresponding class. On the basis of the transformed data set, fuzzy rule learning approaches can be used to measure the fuzzy membership degrees of each instance to each of the predefined class labels. Through cross validation, each of the instances would be used in turn as a test instance to be measured on the extent to which the instance belongs to each single class. Finally, all these instances would have been assigned fuzzy truth values regarding the fuzzy membership degrees to each of the classes, which can be easily discretised to binary truth values.

Based on the above descriptions, we will explore the adoption of multi-task learning approaches in future work for the context of both supervised learning and semi-supervised learning. In addition, we will use multi-task learning approaches for identification of the relationships between different classes such as generalization and aggregation, especially when fuzzy approaches are adopted.

In terms of multi-granularity semi-random data partitioning, we showed in Chap. 3 through experimental results that random partitioning and semi-random partitioning could lead to the same distribution of class frequencies, but different performance of learning. We argued that it could be very likely due to the sample representativeness issue. In order to address this issue, we will propose to divide each class into subclasses. In particular, we adopt rule induction approaches for learning a set of rules from the whole data set, and each of the learned rules would cover a subset of instances, which belong to a subclass of the class assigned to these instances. In this context, each subset of instances, which is covered by a rule learned from the whole data set, is divided into a training subset and a test subset, through random sampling of training/test instances from the subset. Following the partitioning of each subset, all the training/test subsets are merged into a whole training/test set. In this way, it is expected that the training instances are highly similar to the test instances, such that the sample representativeness issue is solved.

In terms of multi-granularity rule learning, we argued that the collaborative rule learning approach presented in Fig. 7.4 can still be advanced through changing the strategies in terms of evaluating attribute-value pairs for appending terms into the left had side of a rule. In particular, the collaborative rule learning approach involves learning a single rule by using each heuristic for the evaluation of attribute-value pairs, and then comparing these rules learned by using different heuristics in terms of rule quality. The above way of using heuristics may lead to the case that some of the selected attribute-value pairs may impact positively on the rule quality, but some others may impact negatively on the quality.

In order to address the above issue, we propose to use all the heuristics in a competitive way towards selecting an attribute-value pair as a rule term. In particular, at each iteration of learning a rule, each heuristic is used for evaluating attribute-value pairs towards selecting the best one as a term of this rule. In this context, the attribute-value pairs selected by different heuristics are compared in terms of the contributions to optimizing the quality of this rule, and then the best one of the attribute-value pairs is finally appended to the left hand side of this rule as a rule term. In this way, it is expected that the quality of each single rule is improved towards classifying accurately unseen instances.

In terms of sentiment analysis through using fuzzy approaches, we suggest the following directions [15] that may lead to advances in text processing:

It is recommended to focus on approaches for multi-granularity processing of textual data (including sentiment data) in the setting of granular computing. In particular, the four questions, which were raised in Chap. 8 regarding how to adopt effectively the multi-granularity framework illustrated in Fig. 8.1, are worth considering towards effective granulation of fuzzy information and effective determination of the number of granularity levels and the number of information granules involved in each level of granularity. The number of granularity levels and the number of information granules involved in each level of granularity do not only impact on the depth of learning of sentiment classifiers but also on the interpretation of classification results. In other words, increasing the above two numbers can increase the depth of learning, but may make it more difficult to interpret the derivation of classification results through a bottom up process (from the bottom level to the top level of granularity), which indicates the importance of effective determination of the above two numbers.

Also, computing with words, which is proposed in [16] and a principal motivation of fuzzy logic, will be explored towards advancing the proposed multi-granularity approach of text processing. In fact, it is much easier to classify words or small phrases than to classify sentences, paragraphs or even documents, especially in the context of polarity classification. In particular, each single word or small phrase can be classified depending on its role and position in a sentence, i.e. some words are more important than other words in a sentence. Following the classification of words/phrases, sentences can be classified through weighted voting of the word classifications, i.e. a sentence can be given a degree of fuzzy membership to the positive/negative class. On this basis, classifying a higher level of instances can be undertaken through the bottom-up aggregation described above.

Furthermore, as mentioned in Chap. 8, clustering may be required towards the reduction of data size through modularizing a textual data set. This would be another direction on investigating different clustering techniques towards effective decomposition of a training set of textual instances into a number of modules (subsets), in order to achieve parallel processing of different modules of the training set for speeding up the process of learning. In the case that new data instances are added into a module of the training set, it is also necessary to consider how to involve incremental learning in order to avoid starting the learning process from the beginning.

In addition, regarding the question raised in Chap. 8: how are the two sets of fuzzy membership degrees combined towards having an overall set of fuzzy membership degrees for each sub-instance in each level of granularity (except for the top and bottom levels)? It is necessary to consider which one of the two operations (conjunction and disjunction) should be taken between the two sets of fuzzy membership degrees, towards having an overall set of fuzzy membership degrees for each sub-instance in a particular level of granularity, i.e. the minimum or the maximum of the two fuzzy membership degrees (from the two sets, respectively) for each class should be taken as the overall degree of fuzzy membership to the class.

References

1. J. Zhang. 1992. Selecting typical instances in instance-based learning. In *Proceedings of the Ninth International Workshop on Machine Learning*, Aberdeen, United Kingdom, 1–3 July 1992, 470–479.
2. Cristianini, N. 2000. *An introduction to support vector machines and other kernel-based learning methods*. Cambridge: Cambridge University Press.
3. Lipowski, A., and D. Lipowska. 2012. Roulette-wheel selection via stochastic acceptance. *Physica A: Statistical Mechanics and its Applications* 391 (6): 2193–2196.
4. Mitchell, T. 1997. *Machine Learning*. New York: McGraw Hill.
5. K. Reynolds, A. Kontostathis, and L. Edwards. 2011. Using machine learning to detect cyberbullying. In *Proceedings of the 10th International Conference on Machine Learning and Applications*, December 2011, 241–244.
6. H. Liu, M. Cocea, A. Mohasseb, and M. Bader. 2017. Transformation of discriminative single-task classification into generative multi-task classification in machine learning context. In *International Conference on Advanced Computational Intelligence*, Doha, Qatar, 4–6 February 2017, 66–73.
7. Zhu, X., and A.B. Goldberg. 2009. *Introduction to semi-supervised learning*. San Rafael: Morgan and Claypool Publishers.
8. Zadeh, L. 2015. Fuzzy logic: A personal perspective. *Fuzzy Sets and Systems* 281: 4–20.
9. Burnap, P., and M. Williams. 2015. Cyber hate speech on twitter: An application of machine classification and statistical modeling for policy and decision making. *Policy and Internet* 7 (2): 223–242.
10. Burnap, P., and M. Williams. 2016. Us and them: identifying cyber hate on twitter across multiple protected characteristics. *EPJ Data Science*, 5(11).
11. Longadge, R., S.S. Dongre, and L. Malik. 2013. Class imbalance problem in data mining: Review. *International Journal of Computer Science and Network* 2 (1): 83–87.
12. Ali, A., S.M. Shamsuddin, and A.L. Ralescu. 2015. Classification with class imbalance problem: A review. *International Journal of Advanced Soft Computing Applications* 7 (3): 176–204.
13. J. Yao. 2005. Information granulation and granular relationships. In *IEEE International Conference on Granular Computing*, Beijing, China, 25–27 July 2005, 326–329.
14. H. Liu and M. Cocea. 2017. Semi-random partitioning of data into training and test sets in granular computing context. *Granular Computing* 2 (4).
15. H. Liu and M. Cocea. Fuzzy information granulation towards interpretable sentiment analysis. *Granular Computing* 3 (1), In press.
16. Zadeh, L. 2002. From computing with numbers to computing with words: From manipulation of measurements to manipulation of perceptions. *International Journal of Applied Mathematics and Computer Science* 12 (3): 307–324.

Appendix A
Results on Generative Multi-task Classification

See Tables A.1, A.2, A.3, A.4.

Table A.1 Results on autos dataset [1]

ID	Class	−1	−2	−3	0	1	2	3	Output
17	0	0.77	0	0	0.73	0.20	0	0	−1
130	0	0	0	0	0.5	0	0	0.84	3
141	0	0	0	0	0.75	0	0.7	0	0
150	1	0	0	0	0.74	0.82	0	0	1
172	2	0	0	0	0.83	0	0.68	0	0
181	−1	0.6	0	0	0.67	0	1	0.5	2
197	−1	1	1	0	0	0	0	0	−1

Table A.2 Results on heart-c dataset [1]

ID	Num	50	50_1	50_2	50_3	50_4	Output
11	50	1	1	0	0	0	50
28	50	1	1	0	0	0	50
36	50_1	1	1	0	0	0	50
82	50	1	0.83	0	0	0	50
102	50	0.8	1	0	0	0	50_1
113	50_1	0.81	1	0	0	0	50_1
138	50_1	0.96	1	0	0	0	50_1
194	50_1	1	1	0	0	0	50
271	50_1	0.98	1	0	0	0	50_1
291	50_1	1	1	0	0	0	50

© Springer International Publishing AG 2018
H. Liu and M. Cocea, *Granular Computing Based Machine Learning*,
Studies in Big Data 35, https://doi.org/10.1007/978-3-319-70058-8

Table A.3 Results on heart-h dataset [1]

ID	Num	50	50_1	50_2	50_3	50_4	Output
17	50	0.9	1	0	0	0	50_1
68	50	0.6	1	0	0	0	50_1
86	50	1	0.9	0	0	0	50
92	50	1	1	0	0	0	50
93	50	1	1	0	0	0	50
96	50	0.74	1	0	0	0	50_1
107	50	0.7	1	0	0	0	50_1
124	50	1	1	0	0	0	50
127	50	1	1	0	0	0	50
160	50	1	1	0	0	0	50
169	50	1	1	0	0	0	50
214	50_1	1	1	0	0	0	50
228	50_1	1	0.6	0	0	0	50
231	50_1	1	1	0	0	0	50
293	50_1	1	1	0	0	0	50

Table A.4 Results on zoo dataset [1]

ID	Type	1	2	3	4	5	6	7	Output
1	6	1	0	0	0	1	1	0	6
2	3	0	0	1	0	1	1	0	6
3	6	1	0	0	0	1	1	0	6
6	6	1	0	0	0	1	1	0	6
9	6	1	0	0	0	1	1	0	6
10	6	1	0	0	0	1	1	0	6
18	3	0	0	1	0	1	1	0	6
21	2	0	1	0	0	0	1	0	6
24	4	0	0	0	1	1	0	0	5
31	6	1	0	0	0	1	1	0	6
36	6	1	0	0	0	1	1	0	6
37	2	0	1	0	0	0	1	0	6
41	2	0	1	0	0	0	1	0	6
42	4	0	0	0	1	1	0	0	5
46	5	0	0	0	1	1	0	0	5
47	6	1	0	0	0	1	1	0	6
67	6	1	0	0	0	1	1	0	6
68	6	1	0	0	0	1	1	0	6
71	2	0	1	0	0	0	1	0	6
73	3	0	0	1	0	1	1	0	6
82	3	0	0	1	0	1	1	0	6
83	2	0	1	0	0	0	1	0	6
84	6	1	0	0	0	1	1	0	6
92	3	0	0	1	0	1	1	0	6
97	4	0	0	0	1	1	0	0	5
100	2	0	1	0	0	0	1	0	6

Reference

1. H. Liu, M. Cocea, A. Mohasseb, and M. Bader. 2017. Transformation of discriminative single-task classification into generative multi-task classification in machine learning context. In *International Conference on Advanced Computational Intelligence*, Doha, Qatar, 4–6 February 2017, 66–73.

Appendix B
Results on Random Data Partitioning

See Figs. B.1, B.2, B.3, B.4, B.5, B.6, B.7, B.8, B.9.

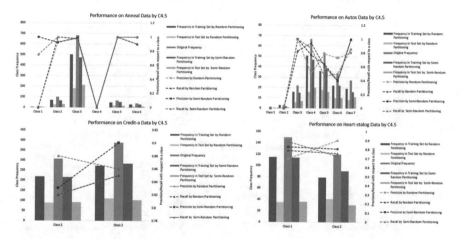

Fig. B.1 Class distribution and performance (precision and recall) by C4.5 for random and semi-random partitioning for the 'anneal', 'autos', 'credit-a' and 'heart-statlog' datasets [1]

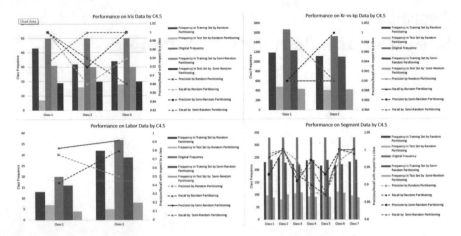

Fig. B.2 Class distribution and performance (precision and recall) by C4.5 for random and semi-random partitioning for the 'iris', 'kr-vs-kp', 'labor' and 'segment' datasets [1]

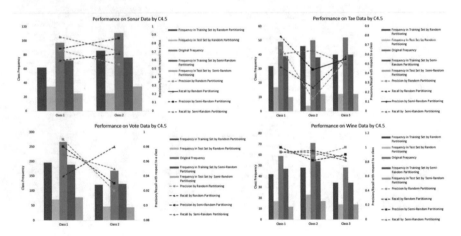

Fig. B.3 Class distribution and performance (precision and recall) by C4.5 for random and semi-random partitioning for the 'sonar', 'tae', 'vote' and 'wine' datasets [1]

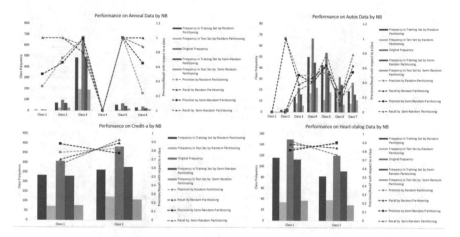

Fig. B.4 Class distribution and performance (precision and recall) by NB for random and semi-random partitioning for the 'anneal', 'autos', 'credit-a' and 'heart-statlog' datasets [1]

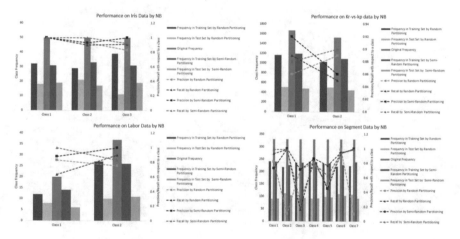

Fig. B.5 Class distribution and performance (precision and recall) by NB for random and semi-random partitioning for the 'iris', 'kr-vs-kp', 'labor' and 'segment' datasets [1]

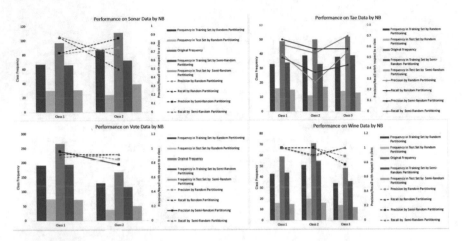

Fig. B.6 Class distribution and performance (precision and recall) by NB for random and semi-random partitioning for the 'sonar', 'tae', 'vote' and 'wine' datasets [1]

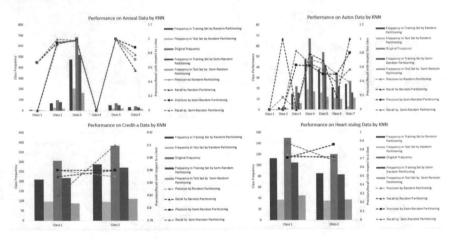

Fig. B.7 Class distribution and performance (precision and recall) by K-NN for random and semi-random partitioning for the 'anneal', 'autos', 'credit-a' and 'heart-statlog' datasets [1]

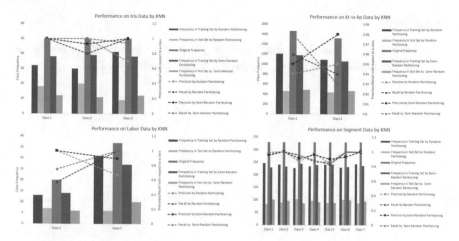

Fig. B.8 Class distribution and performance (precision and recall) by K-NN for random and semi-random partitioning for the 'iris', 'kr-vs-kp', 'labor' and 'segment' datasets [1]

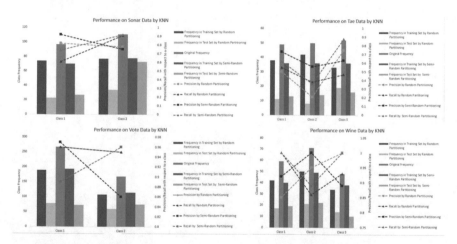

Fig. B.9 Class distribution and performance (precision and recall) by k-NN for random and semi-random partitioning for the 'sonar', 'tae', 'vote' and 'wine' datasets [1]

Reference

1. H. Liu and M. Cocea. 2017. Semi-random partitioning of data into training and test sets in granular computing context. *Granular Computing*, 2(4).

Glossary

Attribute   A property of a data instance. It is also referred to as feature in pattern recognition, variable in mathematics and statistics, column in linear algebra, or field in database and information systems.

Bagging   It is a way of data sampling with replacement, e.g. randomly pick up one item from a bag and then put it back into the bag.

Bias   It is a kind of prediction error originated from heuristics of learning algorithms.

Class   It represents as a collection of instances, which have high similarity to each other.

Class Imbalance   This happens when a sample contains a much higher percentage of instances of one class than the percentage of instances of the other class(es), e.g. a sample contains 90% positive instances and 10% negative instances.

Classification   A data mining task which assigns instances to target categories (referred to as classes). It is also known as categorical prediction, which means that the output is a discrete value.

Classifier   It is a kind of model used for classification/categorical prediction.

Continuous Attribute   It means that the value of this kind of attribute is numerical, such as height.

Data Partitioning   It is to divide a sample into sub-samples.

Data Set   It is also referred to as sample in mathematics and statistics, which is a collection of instances in machine learning.

Data Size   It is also referred to as sample size, which indicates the number of rows/instances in a data set.

Decision Tree   It is a kind of prediction algorithm/model used for classification or regression.

Discrete Attribute   It means that the value of this kind of attributes is categorical, such as gender.

Ensemble Leaning   It is aimed at learning in a collaborative way, which is very similar to group learning by students.

Entropy   It is a measure of uncertainty, i.e. there is no uncertainty when the value of entropy is 0.

© Springer International Publishing AG 2018
H. Liu and M. Cocea, *Granular Computing Based Machine Learning*,
Studies in Big Data 35, https://doi.org/10.1007/978-3-319-70058-8

Feature Extraction   It is a transformation from unstructured data to structured data by extracting a set of features, i.e. each column in a structured data set represents a feature/attribute.

Feature Selection   It is an operation to filter irrelevant features and only keep relevant features for learning algorithms/models.

Fuzzification   It is an operation to transform crisp concepts into fuzzy concepts, e.g. height is transformed into a value of fuzzy membership degree to one of the linguistic terms 'Tall', 'Middle' and 'Short', and its inverse operation is defuzzification.

Fuzzy Logic   It is an extension of binary logic, which reflects a truth value ranged from 0 to 1.

Fuzzy Set   It is a kind of set defined to have fuzzy boundaries (rather than crisp boundaries) in terms of the membership or non-membership of an element to a specific set, i.e. a fuzzy set shows the degree to which an element belongs to the set. An element can belong to different sets with different degrees of membership to each set.

Granular Computing   It is a kind of structured thinking at the philosophical level and also a kind of structured problem solving at the practical level.

Granulation   It is a main operation of granular computing, which involves decomposing a whole into parts.

Granule   It represents a collection of individuals as a unit.

Granularity   It is a multi-level structure of management used to achieve hierarchical organization of granules, e.g. several departments are organized to form a faculty at a university.

Heuristic   It acts as guidance towards effective control of the learning process, e.g. when entropy is used as the heuristic of a learning algorithm, it is judged that a learning task is completed, once the value of entropy becomes 0 at any point of the learning process.

Instance   It is also referred to as object in pattern recognition, data point, vector in mathematics and statistics, row in linear algebra, or tuple in database and information systems.

Label   It is used as a sign to uniquely identify a class of instances, i.e. it assigns a class a unique name.

Labelling   It is an action to provide an instance with a label to show the class to which the instance belongs.

Machine Learning   It essentially means that machines learn knowledge automatically from previously known data and then use the knowledge to predict on unknown data in the future.

Membership Degree   It is a measure of the degree to which an element belongs to a fuzzy set.

Membership Function   It is a function defined for a fuzzy set towards determining the membership degree value of each element in the set.

Overfitting   It is a phenomenon that a model performs well on training data (the data used for learning the model) but poorly on test data (the data that is not used for learning the model).

Random Forest It is a group of decision trees used for classification in the way of group decision making.

Regression It is also known as numerical prediction, which means that the output is a continuous value.

Rule It is used to reflect relationships between inputs and outputs of a system, i.e. casual relationships, where the inputs appear at the left hand side (if part) of the rule and the outputs appear at the right hand side (then part) of the rule.

Rule Antecedent It is the left hand side of a rule.

Rule Base It is a special type of database, which is used particularly for managing rules.

Rule Consequent It is the right hand side of a rule.

Rule Set It is a collection of rules that are used jointly as a whole model for prediction. In classification, it is also referred to as a rule based classifier.

Rule Term It acts as a part of the antecedent of a rule.

Sample Representativeness It means that a sample used for a learning task can reflect almost all the patterns covered by a full population, such that a model learned from the sample can predict accurately on any new instances.

Sentiment Analysis It is also known as opinion mining, which is aimed at identifying the emotions or attitude of people through natural language processing, text analysis and computational linguistics.

Training Set It is a set of instances (data sample) used for machines to learn.

Test Set It is a set of instances (data sample) used to test the learning performance of machines.

Variance It is a kind of prediction errors originated from random processing of data.

Volume It is a measure of the size of data, e.g. how much space is required for storing specific data.

Velocity It is a measure of the speed of data transmission and processing, e.g. how many orders are received each day and how many transactions are completed each day.

Variety It reflects different kinds of data, i.e. structured data (e.g. spreed sheets) and unstructured data (e.g. text, images, audios, videos and signals).

Veracity It is a measure of the degree to which a sample of data can be trusted in business applications.

Variability It is a measure of the degree to which the characteristics of data can be varied, which can lead to different levels of variance of learning performance.